Macmillan/McGraw-Hill • Glencoe

Grade 3

Math Triumphs

Book 1: Number and Operations and Algebra

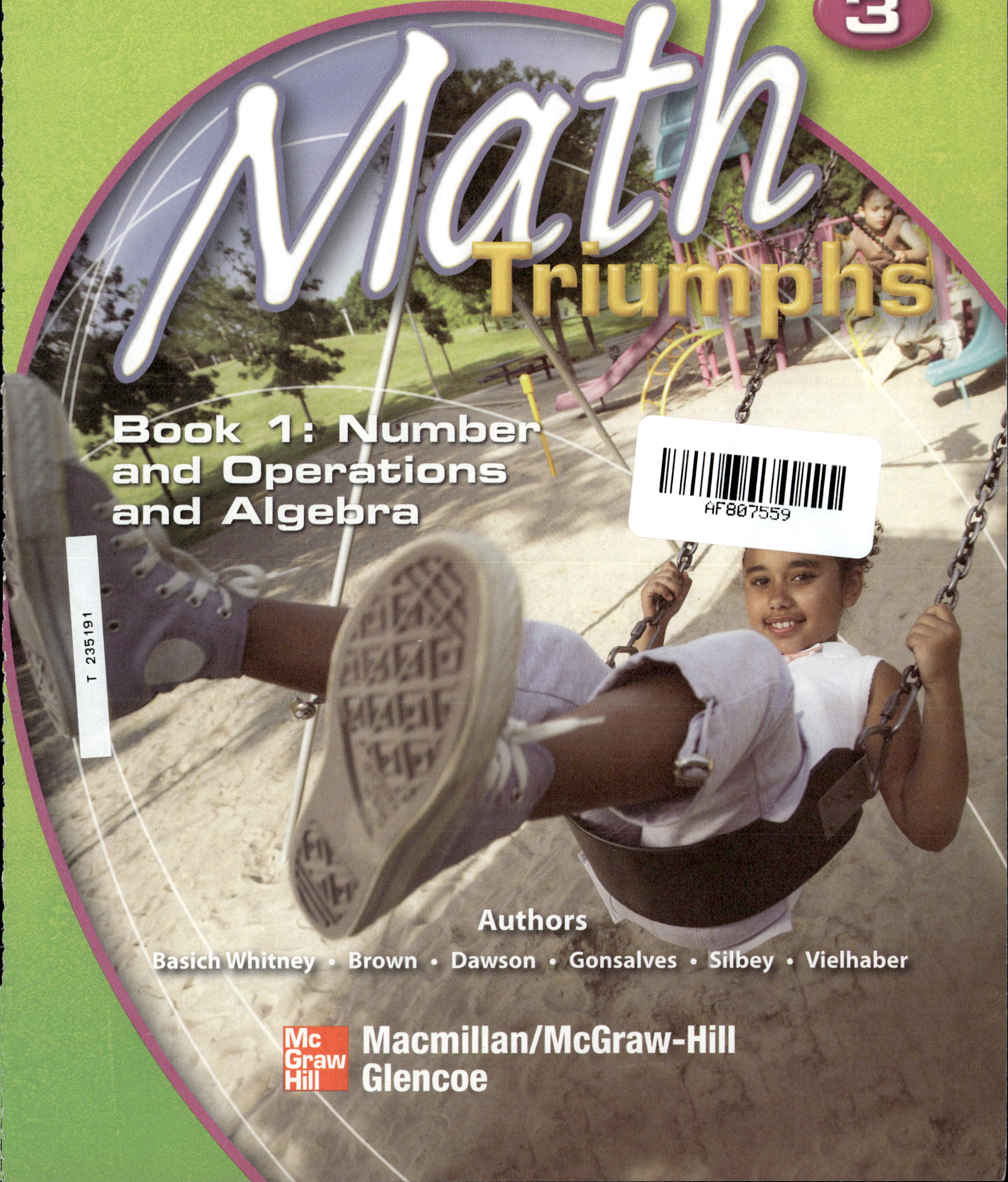

Authors

Basich Whitney • Brown • Dawson • Gonsalves • Silbey • Vielhaber

Mc Graw Hill

Macmillan/McGraw-Hill
Glencoe

Photo Credits

cov Alloy Photography/Veer; **iv** (tl) File Photo, (tc) The McGraw-Hill Companies, (tr) The McGraw-Hill Companies, (cl) Doug Martin, (c) Doug Martin, (cr) Doug Martin, (bl) File Photo, (bc) File Photo; **vi** Richard Hutchings/Digital Light Source; **vii** Richard Hutchings/Digital Light Source; **viii** Ed-Imaging; **1** Richard Hutchings/Digital Light Source; **3** (tl) G.K. Vikki Hart/Getty Images, (tr) G.K. Vikki Hart/Getty Images, (b) Richard Hutchings/Digital Light Source; **5** Getty Images; **6** The McGraw-Hill Companies; **7** McGraw-Hill/Ed-Imaging; **10** Ryan McVay/Getty Images; **12** (tl) Getty Images, (tc) The McGraw-Hill Companies, (tr) C Squared Studios/Getty Images, (bl) Stockdisc/PunchStock, (bcl) Stockdisc/PunchStock, (bc) Getty Images, (br) Burke/Triolo Productions/Getty Images; **13** Richard Hutchings/Digital Light Source; **15** (t) C Squared Studios/Getty Images, (c) Stockbyte/PictureQuest, (b) Getty Images; **17** Rubberball; **18** Getty Images; **20** G.K. Vikki Hart/Getty Images; **21** (t) Stockdisc/PunchStock, (c) Stockdisc/PunchStock, (b) Nikreates/Alamy Images; **27** Brand X Pictures/PunchStock; **29** Richard Hutchings/Digital Light Source; **34** (t) Jeremy Woodhouse/Getty Images, (b) G.K. Vikki Hart/Getty Images; **36** Stockbyte; **37** G.K. Vikki Hart/Getty Images; **40** (l) Kevin Peterson/Getty Images, (tr) Ed-Imaging, (br) Stockbyte; **43** Richard Hutchings/Digital Light Source; **45** (t) The McGraw-Hill Companies, (c) The McGraw-Hill Companies, (b) Richard Hutchings/Digital Light Source; **46** Getty Images; **47** (t) Ryan McVay/Getty Images, (c) Daniel Templeton/Alamy Images, (b) Getty Images; **48** (t) CORBIS, (c) C Squared Studios/Getty Images, (b) C Squared Studios/Getty Images; **49** Ed-Imaging; **55** Richard Hutchings/Digital Light Source; **57** (l) Getty Images, (tr) C Squared Studios/Getty Images, (br) Burke/Triolo Productions/Getty Images; **59** Steve Ohlsen/Alamy Images; **63** (l) The McGraw-Hill Companies, (r) Brand X Pictures/PunchStock; 65 F. Schussler/Getty Images; **67** Getty Images; **69** Ed-Imaging; **71** Brand X Pictures/PunchStock; **75** Getty Images; **76** (t) Getty Images, (b) James Urbach/SuperStock; **78** Ed-Imaging.; **81** Richard Hutchings/Digital Light Source; **84** (t) Getty Images, (b) Stockdisc/PunchStock; **85** C Squared Studios/Getty Images; **86** Getty Images; **87** Masterfile; **89** Siede Preis/Getty Images; **90** McGraw Hill Companies/Ken Canavaugh, Photographer; **91** (t) Siede Preis/Getty Images, (b) Creatas/PunchStock; **93** Ed-Imaging; **95** (tl) Stockdisc/PunchStock, (tr) Stockdisc/PunchStock, (b) Getty Images; **96** Stockdisc/PunchStock; **97** Image Source/PunchStock; **98** D. Hurst/Alamy Images; **100** C Squared Studios/Getty Images; **101** C Squared Studios/Getty Images; **102** (tl) Getty Images, (all others) G.K. Vikki Hart/Getty Images; **103** isifa Image Service s.r.o./Alamy Images; **104** (t) Stockbyte/PunchStock, (b) Siede Preis/Getty Images; **107** Richard Hutchings/Digital Light Source; **109** Getty Images; **114** Siede Preis/Getty Images; **116** (tl) Ed-Imaging, (tr) Randy Faris/CORBIS, (c) C Squared Studios/Getty Images, (b) C Squared Studios/Getty Images.

The McGraw-Hill Companies

Macmillan/McGraw-Hill
Glencoe

Send all inquiries to:
Macmillan/McGraw-Hill • Glencoe/McGraw-Hill
8787 Orion Place
Columbus, OH 43240-4027

ISBN: 978-0-07-888198-5
MHID: 0-07-888198-6

Printed in the United States of America.

3 4 5 6 7 8 9 10 066 16 15 14 13 12 11 10 09 08

Math Triumphs
Grade 3, Book 1

Book 1

Chapter 1 Addition and Subtraction

Chapter 2 Introduction to Multiplication

Chapter 3 Introduction to Division

Book 2

Chapter 4 Place Value

Chapter 5 Fractions

Chapter 6 Fraction Equivalence

Book 3

Chapter 7 Geometry

Chapter 8 Measurement and Geometry

Chapter 9 Data Analysis

Authors and Consultants

CONSULTING AUTHORS

Frances Basich Whitney
Project Director, Mathematics K–12
Santa Cruz County Office of Education
Capitola, California

Kathleen M. Brown
Math Curriculum Staff Developer
Washington Middle School
Long Beach, California

Dixie Dawson
Math Curriculum Leader
Long Beach Unified
Long Beach, California

Philip Gonsalves
Mathematics Coordinator
Alameda County Office of Education
Hayward, California

Robyn Silbey
Math Specialist
Montgomery County Public Schools
Gaithersburg, Maryland

Kathy Vielhaber
Mathematics Consultant
St. Louis, Missouri

CONTRIBUTING AUTHORS

Viken Hovsepian
Professor of Mathematics
Rio Hondo College
Whittier, California

FOLDABLES Study Organizer

Dinah Zike
Educational Consultant
Dinah-Might Activities, Inc.
San Antonio, Texas

CONSULTANTS

Assessment

Donna M. Kopenski, Ed.D.
Math Coordinator K–5
City Heights Educational Collaborative
San Diego, California

Instructional Planning and Support

Beatrice Luchin
Mathematics Consultant
League City, Texas

ELL Support and Vocabulary

ReLeah Cossett Lent
Author/Educational Consultant
Alford, Florida

Reviewers

Each person below reviewed at least two chapters of the Student Study Guide, providing feedback and suggestions for improving the effectiveness of the mathematics instruction.

Dana M. Addis
Teacher Leader
Dearborn Public Schools
Dearborn, MI

Renee M. Blanchard
Elementary Math Facilitator
Erie School District
Erie, PA

Jeanette Collins Cantrell
5th and 6th Grade Math Teacher
W.R. Castle Memorial Elementary
Wittensville, KY

Helen L. Cheek
K-5 Math Specialist
Durham Public Schools
Durham, NC

Mercy Cosper
1st Grade Teacher
Pershing Park Elementary
Killeen, TX

Bonnie H. Ennis
Math Coordinator
Wicomico County Public Schools
Salisbury, MD

Sheila A. Evans
Instructional Support Teacher - Math
Glenmount Elementary/Middle School
Baltimore, MD

Lisa B. Golub
Curriculum Resource Teacher
Millennia Elementary
Orlando, FL

Donna Hagan
Program Specialist - Special Programs Department
Weatherford ISD
Weatherford, TX

Russell Hinson
Teacher
Belleview Elementary
Rock Hill, SC

Tania Shepherd Holbrook
Teacher
Central Elementary School
Paintsville, KY

Stephanie J. Howard
3rd Grade Teacher
Preston Smith Elementary
Lubbock, TX

Rhonda T. Inskeep
Math Support Teacher
Stevens Forest Elementary School
Columbia, MD

Albert Gregory Knights
Teacher/4th Grade/Math Lead Teacher
Cornelius Elementary
Houston, TX

Barbara Langley
Math/Science Coach
Poinciana Elementary School
Kissimmee, FL

David Ennis McBroom
Math/Science Facilitator
John Motley Morehead Elementary
Charlotte, NC

Jan Mercer, MA; NBCT
K-5 Math Lab Facilitator
Meadow Woods Elementary
Orlando, FL

Rosalind R. Mohamed
Instructional Support Teacher - Math
Furley Elementary School
Baltimore, MD

Patricia Penafiel
Teacher
Phyllis Miller Elementary
Miami, FL

Lindsey R. Petlak
2nd Grade Instructor
Prairieview Elementary School
Hainesville, IL

Lana A. Prichard
District Math Resource Teacher K-8
Lawrence Co. School District
Louisa, KY

Stacy L. Riggle
3rd Grade Spanish Magnet Teacher
Phillips Elementary
Pittsburgh, PA

Wendy Scheleur
5th Grade Teacher
Piney Orchard Elementary
Odenton, MD

Stacey L. Shapiro
Teacher
Zilker Elementary
Austin, TX

Kim Wilkerson Smith
4th Grade Teacher
Casey Elementary School
Austin, TX

Wyolonda M. Smith, NBCT
4th Grade Teacher
Pilot Elementary School
Greensboro, NC

Kristen M. Stone
3rd Grade Teacher
Tanglewood Elementary
Lumberton, NC

Jamie M. Williams
Math Specialist
New York Mills Union Free School District
New York Mills, NY

Contents

CHAPTER 1 Addition and Subtraction

CHAPTER 2 Introduction to Multiplication

Contents

Chapter 3 Introduction to Division

CHAPTER 1 Addition and Subtraction

Home Connection

English

Dear Family,

Today our class started **Chapter 1, Addition and Subtraction Facts.** In this chapter, I will learn addition and subtraction facts through 18. I will also learn to add and subtract two-digit numbers.

Love, ______________

Spanish

Estimada familia:

Hoy en clase comenzamos **el Capítulo 1, titulado Operaciones de suma y la resta.** En este capítulo aprenderé a sumar y a restar hasta el número 18. También aprenderé a sumar y a restar números de dos dígitos.

Cariños, ______________

Help at Home

You can help your child succeed with addition and subtraction using common household items. Give your child two groups of items and have him or her find how many items there are in all. Or, give your child a group of items and ask him or her how many are left when a certain number of items are taken away.

Math Online Take the chapter Get Ready quiz at macmillanmh.com.

Ayude en casa

Usted puede ayudar a su hijo(a) a dominar la suma y la resta usando objetos comunes del hogar. De a su hijo(a) dos grupos de objetos y pídale que le diga cuántos hay en total. O bien, déle un grupo de objetos y pregúntele cuántos quedan cuando un cierto número de objetos se retira del grupo.

Name ______________________________

Get Ready

Count. Write the number.

1

______

2

______

Circle the greater number.

3 4   7

4 8   2

5 Use the number line. Start at 1 and count on 2 more. Circle the number.

0 1 2 3 4 5 6 7 8 9 10

6 Use the number line. Start at 2 and count on 4 more. Circle the number.

0 1 2 3 4 5 6 7 8 9 10

7 Juan plays soccer. He drew an X for each goal he scored. How many goals did Juan score?

X X X

Juan scored ______ goals.

STOP

Lesson 1-1

Name ______________________________

Addition Facts 0 to 10

Key Concept

Look at the cats.

3 + 2

How many cats are there in all?

3 + 2 = 5

There are 5 cats in all.

Vocabulary

add (addition) join sets together to find the total or sum

2 + 5 = 7

addition sentence a math sentence that has an addition sign in it

4 + 5 = 9

sum the answer to an addition problem

2 + 4 = 6 ⟵ sum

When two groups are added together, they combine to form one larger group.

Example

How many soccer balls are there in all?

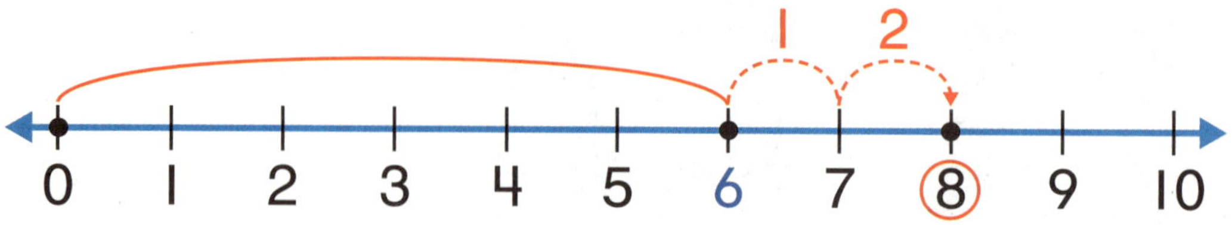

Step 1 Count the soccer balls in the first group. 6

Step 2 Count the soccer balls in the second group. 2

Step 3 Add 6 and 2. $6 + 2 = 8$

Answer There are 8 soccer balls in all.

Step-by-Step Practice

How many footballs are there in all?

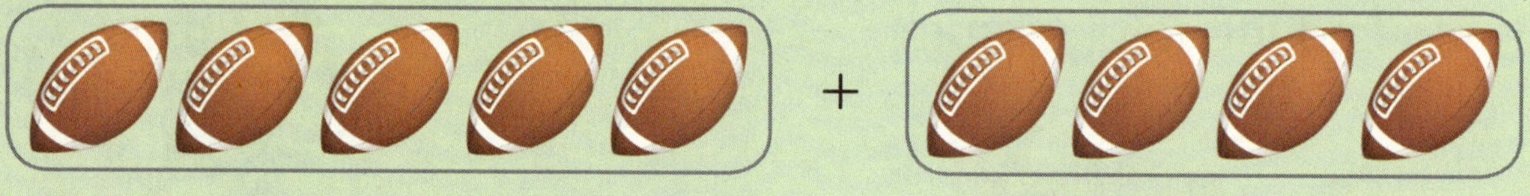

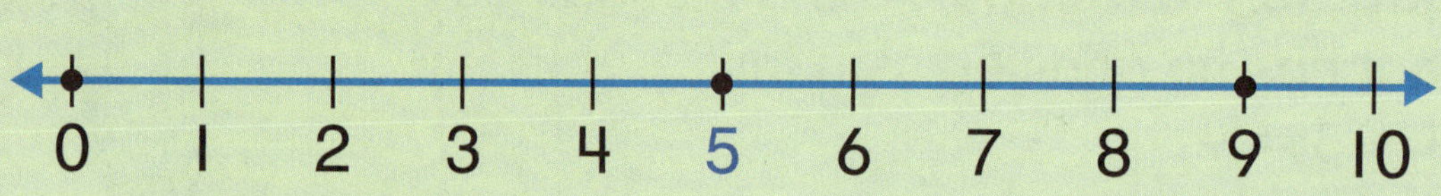

Step 1 Count the footballs in the first group. 5

Step 2 Count the footballs in the second group. ____

Step 3 Add ____ and ____. ____ + ____ = ____

Answer There are ____ footballs in all.

Name ________________________________

Guided Practice

1 Find the sum.

____ + ____ = ____ piñatas in all

Problem-Solving Practice

2 Carmen has 5 blue balloons for her birthday party. Then she blows up 3 more red balloons. How many balloons does Carmen have in all?

Understand Underline key words.

Plan Draw a picture.

Solve Draw ____ circles in the blue box.

Draw ____ circles in the red box. Count the circles.

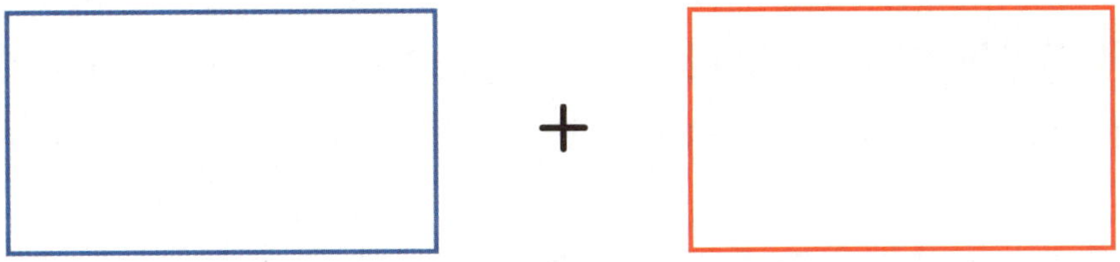

Carmen has ____ balloons in all.

Check Use counters.

GO on

Practice on Your Own

Write an addition sentence. Find each sum.

3 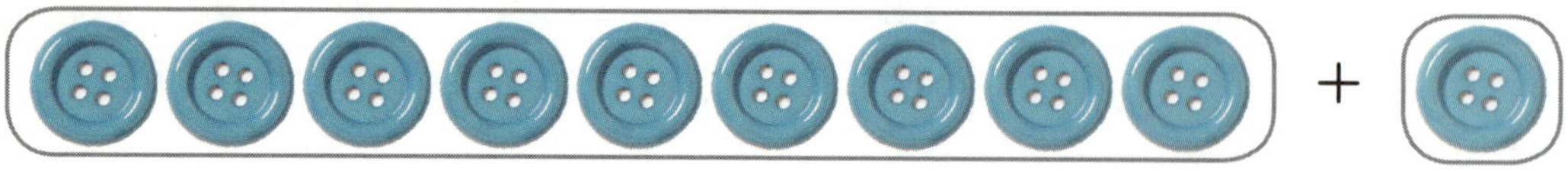

____ + ____ = ____

4 

____ + ____ = ____

Find each sum. Use the number line.

5 8 + 2 = ____

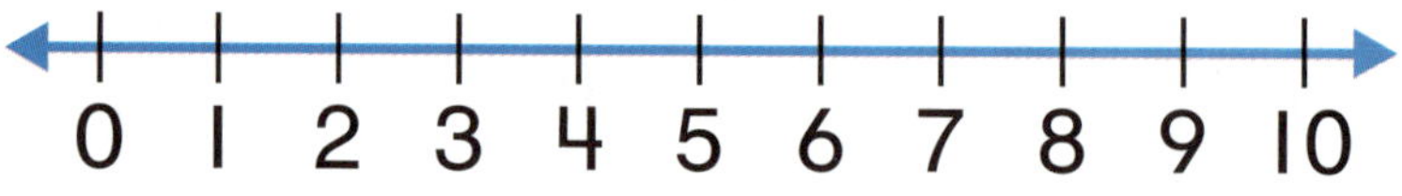

6 3 + 3 = ____

Find each sum.

7 5 + 2 = ____ **8** 1 + 7 = ____ **9** 6 + 4 = ____

10 WRITING IN MATH Julie sees 4 cars. Then she sees 3 more cars. How many cars does Julie see in all? Explain.

____

____

Vocabulary Check Complete.

11 Look at 7 + 1 = 8. The number 8 is the ________.

STOP

Name ______________________

Addition Facts through 18

Key Concept

Look at the cubes.

6
+ 6
―――― ← This line means equals.

How many cubes are there in all?

6 + 6 = 12

There are 12 cubes in all.

Vocabulary

add (addition) join sets together to find the total or sum

2 + 5 = 7

addition sentence a math sentence that has an addition sign in it

4 + 5 = 9

sum the answer to an addition problem

2 + 4 = 6 ← sum

17 is the same as 1 ten and 7 ones.

Example

Find the sum.

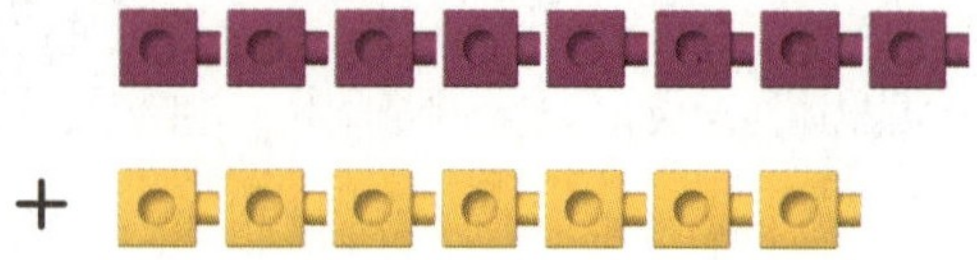

$$\begin{array}{r} 8 \\ +\,7 \\ \hline \end{array}$$

Step 1 Count the first group. 8 cubes

Step 2 Count the second group. 7 cubes

Step 3 Add 8 and 7.

Step 4 Write the addition sentence. 8 + 7 = 15

Answer There are 15 connecting cubes.

Step-by-Step Practice

Find the sum.

Mercedes has 9 baseball cards. She buys 5 more. How many baseball cards does Mercedes have in all?

 + 

9 + ____

Step 1 Count the first group. 9 cards

Step 2 Count the second group. ____ cards

Step 3 Add ____ and ____.

Step 4 Write the addition sentence.

____ + ____ = ____

Answer Mercedes has ____ baseball cards in all.

Name ______________________________

Guided Practice

Find each sum.

1 ____ + ____ = ____

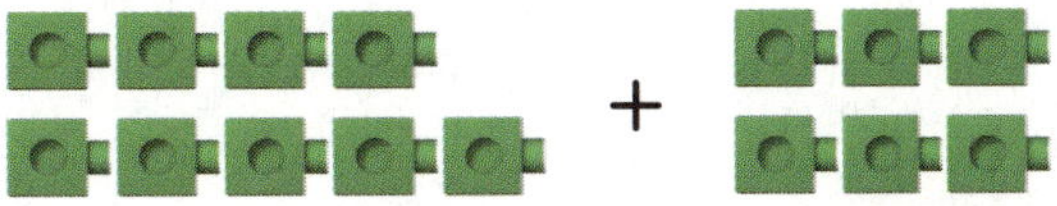

2
$$\begin{array}{r} 9 \\ +\ 9 \\ \hline \square \end{array}$$

3
$$\begin{array}{r} 7 \\ +\ 6 \\ \hline \square \end{array}$$

4
$$\begin{array}{r} 4 \\ +\ 7 \\ \hline \square \end{array}$$

Problem-Solving Practice

5 Catherine has 6 blue crayons and 5 red crayons.
How many crayons does she have in all?

Understand Underline key words.

Plan Draw a picture.

Solve Draw ____ blue crayons in the blue box.

Draw ____ red crayons in the red box.

☐ + ☐ Count the crayons.

Catherine has ____ crayons in all.

Check Use a number line.

GO on

Practice on Your Own

Find each sum.

6

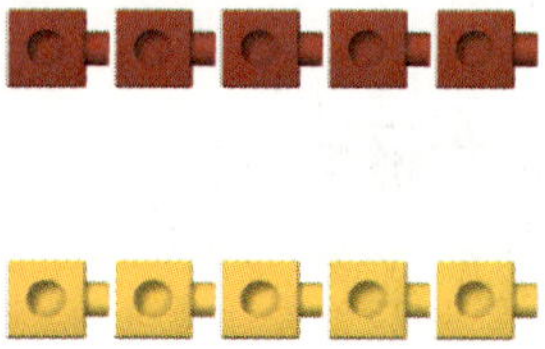

$+$ □ / □ = □

7

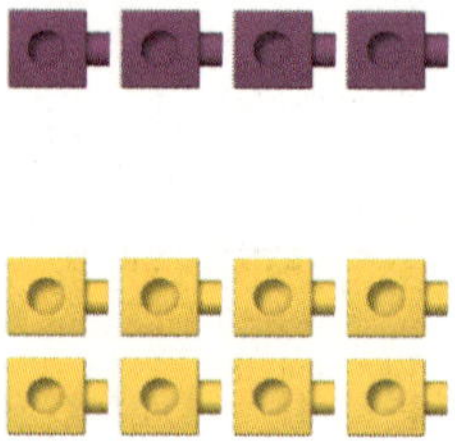

$+$ □ / □ = □

8

____ + ____ = ____

Add.

9	10	11	12
3 + 9 □	8 + 5 □	6 + 9 □	9 + 4 □

13 **WRITING IN MATH** Omar has 18 flowers in his garden. How can you show 18 using tens and ones? Explain.

____

____

____

Vocabulary Check Complete.

14 The addition sign (+) is used in an ____________________

____________________.

STOP

Name ____________________

Progress Check 1 (Lessons 1-1 and 1-2)

1 Billy saw seagulls at the beach. Write an addition sentence to show how many seagulls he saw in all.

____ + ____ = ____

Find each sum.

2 3 + 7 = ____

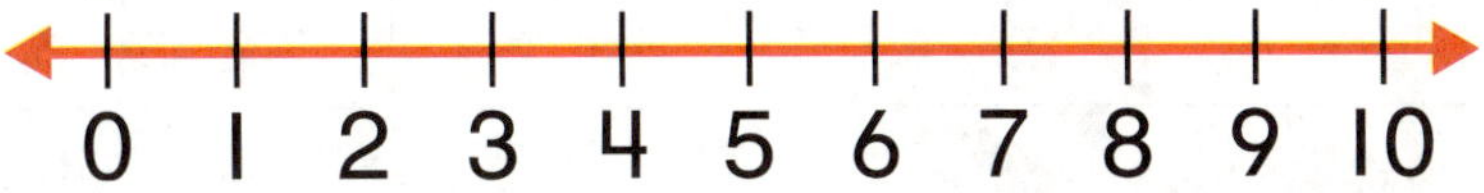

3
$$\begin{array}{r} 5 \\ +\ 9 \\ \hline \square \end{array}$$

4
$$\begin{array}{r} 6 \\ +\ 7 \\ \hline \square \end{array}$$

5
$$\begin{array}{r} 9 \\ +\ 8 \\ \hline \square \end{array}$$

---

6 Isabel found 6 seashells at the beach yesterday.
She found 2 seashells at the beach today.
How many seashells did Isabel find in all?

Isabel found ______ seashells in all.

Name ________________________________

«« Replay Number Picnic

Solve the problems to find your way through the picnic area. If a square has a sum equal to 14 or greater, draw an X on it.

Start

8 + 8 = ☐	2 + 9 = ☐	6 + 1 = ☐	4 + 9 = ☐
9 + 8 = ☐	7 + 7 = ☐	7 + 4 = ☐	5 + 7 = ☐
6 + 7 = ☐	8 + 9 = ☐	5 + 8 = ☐	8 + 1 = ☐
3 + 5 = ☐	9 + 7 = ☐	9 + 9 = ☐	8 + 6 = ☐

Finish

Lesson 1-3

Name ______________________

Subtraction Facts 0 to 10

Key Concept

Look at the beach balls.
There were 7. Then 5 were crossed out.

7 − 5

How many beach balls are left?
7 − 5 = 2
There are 2 beach balls left.

Vocabulary

subtract (subtraction) take away, take apart, separate, or find the difference between two sets
5 − 5 = 0

subtraction sentence a math sentence that has a subtraction sign in it 9 − 4 = 5

difference the answer to a subtraction problem 3 − 1 = 2 ⟵ The difference is 2.

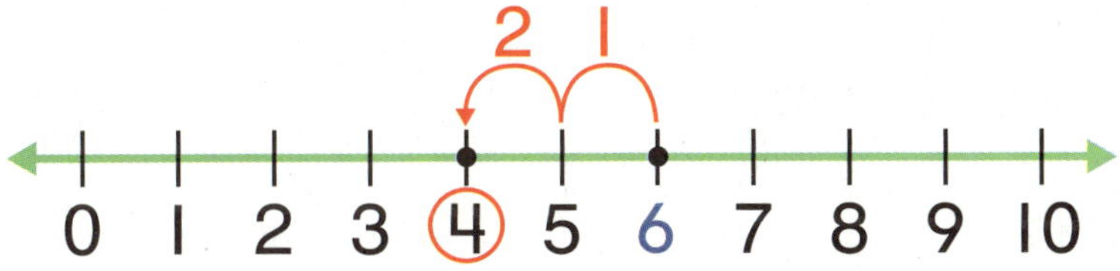

You can also use a number line to subtract.

Example

Six birds fly away. How many birds are left?

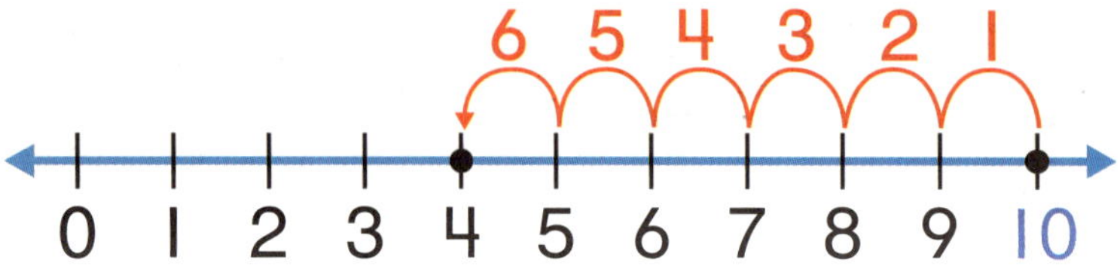

Step 1 Count the birds. There are 10 birds.

Step 2 Start at 10. Count back 6.

Step 3 Write the subtraction sentence. $10 - 6 = 4$

Answer There are 4 birds left.

Step-by-Step Practice

Seven ducks swim away. How many ducks are left?

Step 1 Count the ducks. There are __9__ ducks.

Step 2 Start at ____. Count back 7.

Step 3 Write the subtraction sentence.

____ – ____ = ____

Answer There are ____ ducks left.

Name ____________________

Guided Practice

Find each difference.

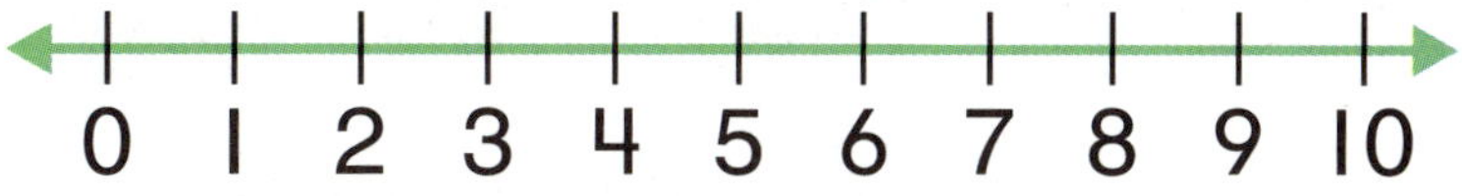

1  ____ − ____ = ____

2 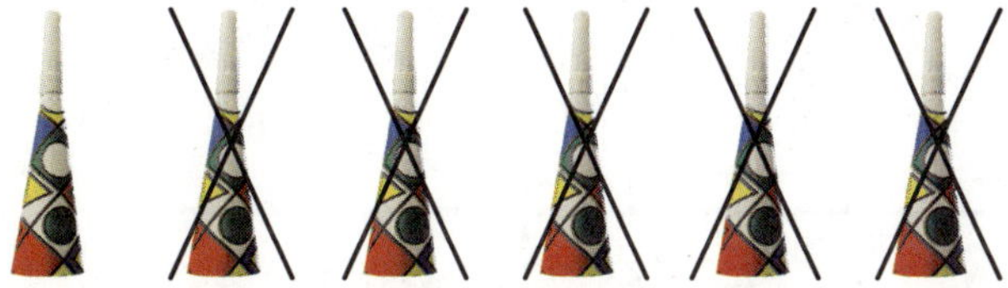 ____ − ____ = ____

Problem-Solving Practice

3 Kelly receives 9 birthday presents. She opens 5 of the presents. How many presents does Kelly have left to open?

Understand Underline key words.

Plan Use a number line.

Solve Start at ____. Count back ____.
Circle the number.

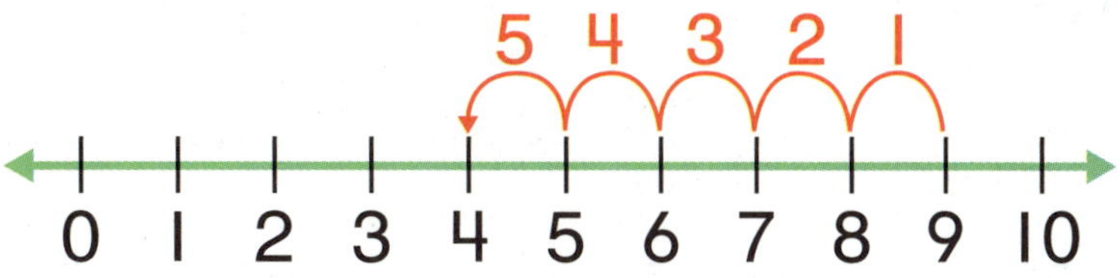

Kelly has ____ presents left to open.

Check Draw a picture.

GO on

Practice on Your Own

Find each difference.

4 5 − 2 = ____

5 6 − 3 = ____

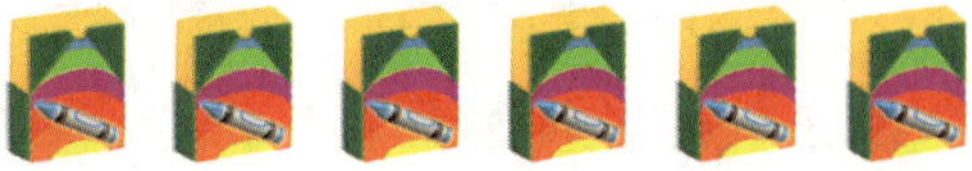

Find each difference. Use the number line.

6 9 − 9 = ____

7 8 − 2 = ____

Find each difference.

8 8 − 4 = ____

9 10 − 2 = ____

10 4 − 0 = ____

11 WRITING IN MATH Mika counts 7 dragonflies. Then 6 dragonflies fly away. How many dragonflies are left? How do you know?

____

____

____

Vocabulary Check Complete.

12 The answer to a subtraction problem is the ____________.

STOP

Lesson 1-4

Name ______________________________

Subtraction Facts through 18

Key Concept

Look at the cubes.

There were 11. Then 5 were crossed out.

How many cubes are left?

$11 - 5 = 6$

There are 6 cubes left.

$$\begin{array}{r} 11 \\ -\ 5 \\ \hline 6 \end{array}$$

← This line means equals.

Vocabulary

subtract (subtraction) take away, take apart, separate, or find the difference between two sets

$5 - 5 = 0$

subtraction sentence a math sentence that has a subtraction sign in it

$9 - 4 = 5$

difference the answer to a subtraction problem

$3 - 1 = 2$ ← difference

$10 - 6 = 4$

$4 + 6 = 10$

Subtraction and addition are opposites. Check subtraction with addition.

Example

Find the difference.

Step 1 Count the sand dollars. There are 16 sand dollars.

Step 2 How many sand dollars are crossed out? 4

Step 3 Write a subtraction sentence. $16 - 4 = 12$

Answer The difference is 12.

Step-by-Step Practice

Find the difference.

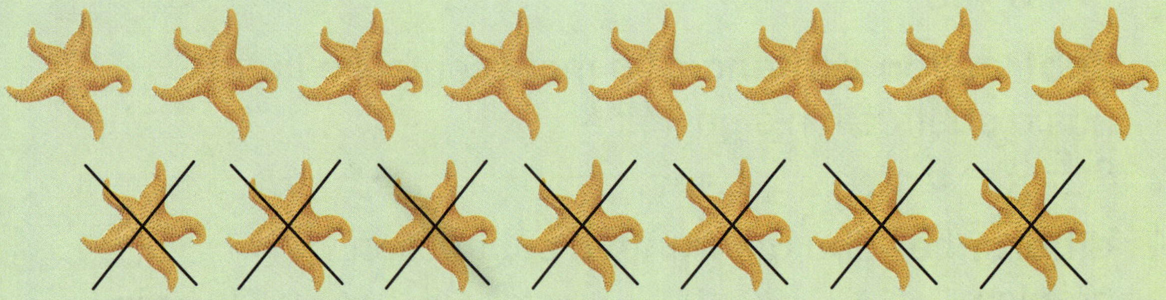

Step 1 Count the starfish. There are 15 starfish.

Step 2 How many starfish are crossed out? ____

Step 3 Write a subtraction sentence.

____ − ____ = ____

Answer The difference is ____.

Name ______________________________

Guided Practice

Find each difference.

1 ____ − ____ = ____

2

$$\begin{array}{r} 18 \\ -\ 9 \\ \hline \square \end{array}$$

3

$$\begin{array}{r} 11 \\ -\ 6 \\ \hline \square \end{array}$$

4

$$\begin{array}{r} 12 \\ -\ 5 \\ \hline \square \end{array}$$

Problem-Solving Practice

5 Allen needs to make 15 foul shots. He has made 9 foul shots. How many more shots does Allen need?

Understand Underline key words.

Plan Use a number line to subtract.

Solve Start at ____. Count back ____. Circle the answer.

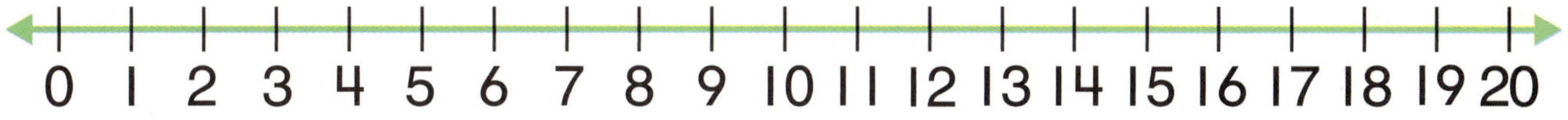

____ − ____ = ____

He needs to make ____ more.

Check Use addition to check your answer.

GO on

Practice on Your Own

Find each difference.

6

13 − 6 = ____

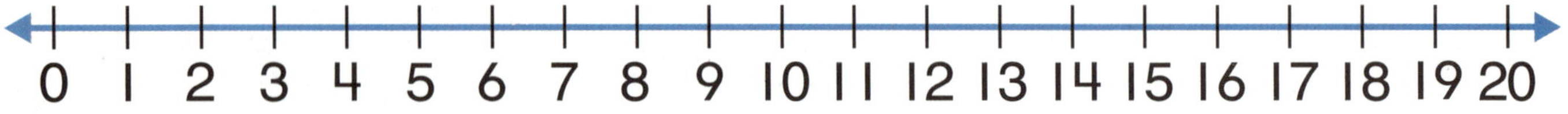

7 12 − 7 = ____

8 17 − 8 = ____

9 14 − 6 = ____

10

$$\begin{array}{r} 16 \\ -\ 9 \\ \hline \square \end{array}$$

11

$$\begin{array}{r} 13 \\ -\ 8 \\ \hline \square \end{array}$$

12

$$\begin{array}{r} 12 \\ -\ 8 \\ \hline \square \end{array}$$

13 **WRITING IN MATH** Ian drew 13 stars. He colored 8 stars yellow. How many stars are not yellow? Explain.

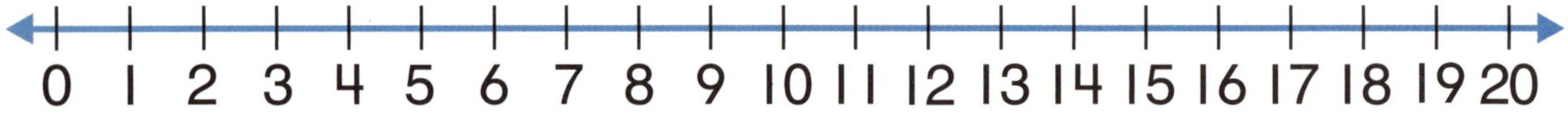

____

____

____

Vocabulary Check Complete.

14 When you take away or find the difference between two sets, you ________________.

STOP

Name ______________________

Progress Check 2 (Lessons 1-3 and 1-4)

Find each difference.

1

____ − ____ = ____

2 9 − 6 = ____

3 10 − 2 = ____

4

$$\begin{array}{r} 15 \\ -\ \ 6 \\ \hline \square \end{array}$$

5

$$\begin{array}{r} 14 \\ -\ \ 3 \\ \hline \square \end{array}$$

6

$$\begin{array}{r} 11 \\ -\ \ 8 \\ \hline \square \end{array}$$

7 Rebecca bought 17 oranges to make orange juice.
She has squeezed 9 oranges so far.
How many oranges are left?

____ − ____ = ____ oranges are left.

8 Mr. Nguyen bought juice boxes for 18 students.
Only 7 students wanted juice.
How many students did not want juice?

______ students did not want juice.

Name ______________________________

«« Replay Play Tic-Tac-Toe!

Find each difference.
Mark an X in the box if the difference is less than 7.
Mark an O in the box if the difference is greater than 7.
Shade the boxes that make Tic-Tac-Toe!

17 − 8 = ____	14 − 5 = ☐	9 − 5 = ____
10 − 5 = ☐	9 − 1 = ____	8 − 2 = ☐
16 − 7 = ____	13 − 9 = ☐	11 − 9 = ____

Lesson 1-5

Name ______________________________

Add Two-Digit Numbers

Key Concept

Look at the crayons. How many crayons are there in all?

41 + 26

	tens	ones
	4	1
+	2	6
	6	7

How many **ones**? 7 ones
How many **tens**? 6 tens
41 + 26 = 67
There are 67 crayons in all.

Vocabulary

tens a place value of a number
The number 23 has 2 tens.

ones a place value of a number
The number 23 has 3 ones.

regroup take apart a number to write it in a new way
1 ten + 2 ones becomes 12 ones.

You can regroup ones into tens and ones.
For example, 17 ones = 1 ten and 7 ones.

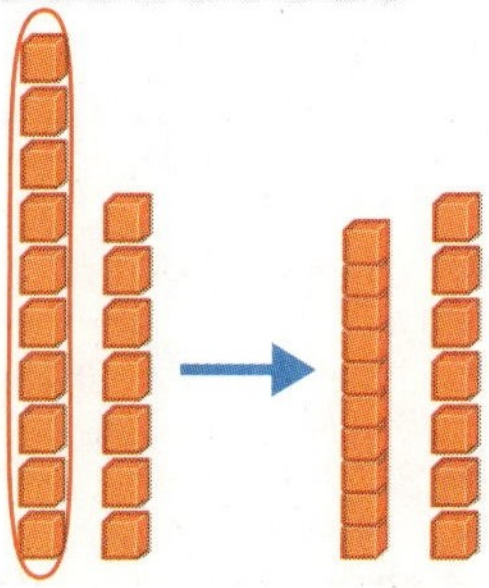

Example 1

Find 43 + 24.

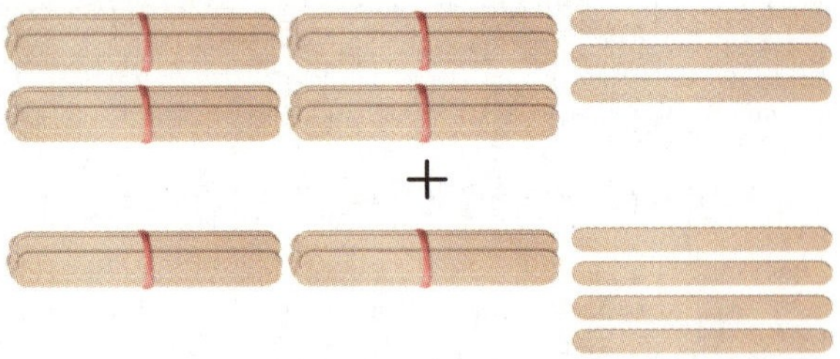

Step 1 Count the number of ones. 7 ones
Write 7 in the ones place.

Step 2 Count the number of tens. 6 tens
Write 6 in the tens place.

	tens	ones
	4	3
+	2	4
	6	7

Answer 43 + 24 = 67

Step-by-Step Practice

Find 53 + 45.

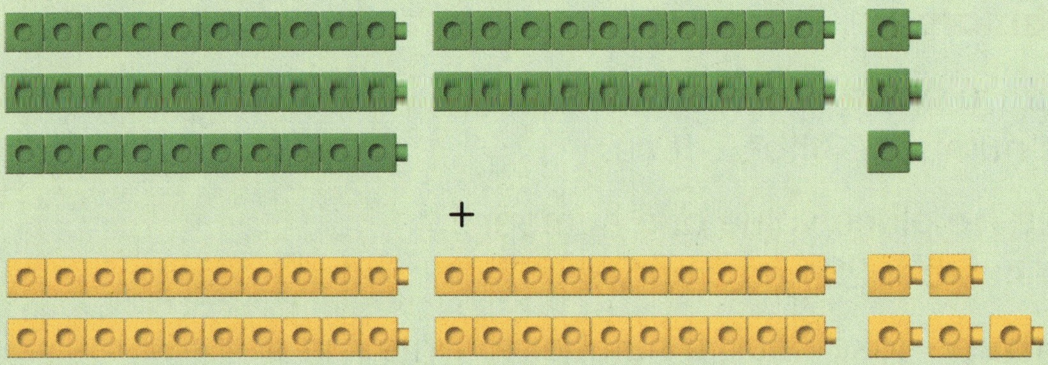

Step 1 Count the number of ones. _____ ones

Write _____ in the ones place.

Step 2 Count the number of tens. _____ tens

Write _____ in the tens place.

	tens	ones
	5	3
+	4	5

Answer _____ + _____ = _____

Example 2

Find 23 + 49.

Step 1 Add the ones column. 3 + 9 = 12

Step 2 Regroup 12 as 1 ten and 2 ones.

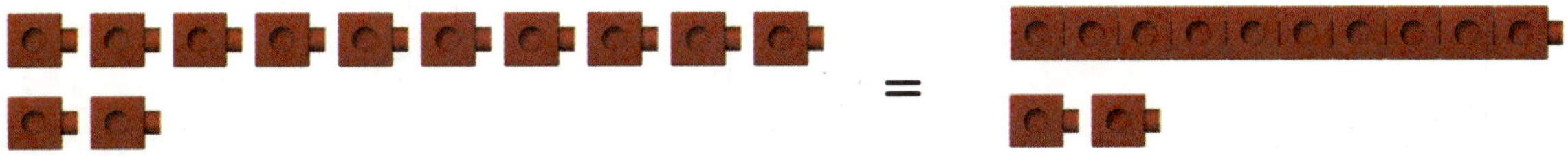

Step 3 Write 2 in the ones place.
Write 1 above the tens column.

Step 4 Add the tens column. 1 + 2 + 4 = 7
Write this sum in the tens place.

Answer 23 + 49 = 72

	tens	ones
	1	
	2	3
+	4	9
	7	2

Step-by-Step Practice

Find 68 + 25.

Step 1 Add the ones column. 8 + 5 = 13

Step 2 Regroup 13 as ____ ten and ____ ones.

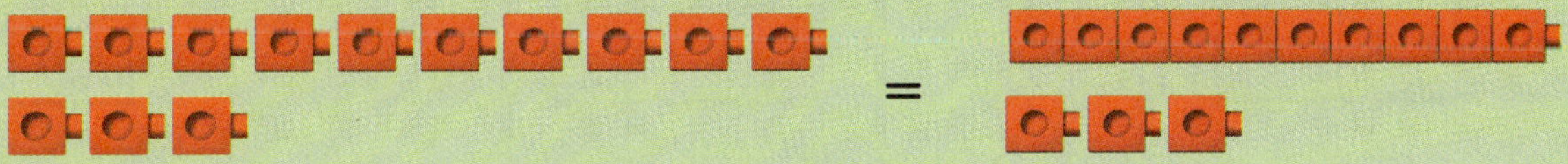

Step 3 Write ____ in the ones place.

Write ____ above the tens column.

Step 4 Add the tens column.

____ + 6 + 2 = ____
Write this sum in the tens place.

Answer 68 + 25 = ____

	tens	ones
	□	
	6	8
+	2	5

Name ________________________________

Guided Practice

Find each sum.

1 16 + 23

	tens	ones
	1	6
+	2	3

2 43 + 11

	tens	ones
	4	3
+		

3 32 + 46

	tens	ones
+		

4 64 + 27

	tens	ones
	□	
	6	4
+	2	7

5 59 + 26

	tens	ones
	□	
	5	9
+		

6 27 + 17

	tens	ones
	□	
+		

7

	□	
	5	4
+	2	7

8

  18
+ 51
□

9

  53
+ 25
□

Problem-Solving Practice

10 Michelle has 58 stickers. She buys 18 more stickers. How many stickers does Michelle have?

Understand Underline key words.

Plan Use a place-value chart.

Solve

□	
+	

Michelle has ________ stickers.

Check Use a hundred chart. Start at 56. Count on 18.

Practice on Your Own

Find each sum.

11 24 + 62

	tens	ones
	2	4
+		

12 37 + 46

	tens	ones
	□	
+		

13 71 + 19

	□	
	7	1
+	1	9

GO on

14 Lilla has 33 acorns. She finds 12 more. How many acorns does Lilla have?

Lilla has ______ acorns.

15 In the first half of a basketball game, Lin's team scored 37 points. In the second half, they scored 49 points. How many points did Lin's team score in all?

Lin's team scored ______ points in all.

16 WRITING IN ▸MATH LaToya solved the following problem, 46 + 38. Is LaToya correct? Solve and explain.

	4	6
+	3	8
	7	14

____

____

Vocabulary Check Complete.

Look at the number.

27

17 The 7 is in the __________ place.

18 The 2 is in the __________ place.

STOP

Lesson 1-6

Name ______________________

Subtract Two-Digit Numbers

Key Concept

Look at the crayons. There were 35. Then 12 were crossed out. How many are left?

$$\begin{array}{r} 35 \\ -\ 12 \\ \hline \end{array}$$

There are 3 **ones** left.
There are 2 **tens** left.

tens	ones
2	3

Vocabulary

tens a place value of a number
The number 23 has 2 tens.

ones a place value of a number
The number 23 has 3 ones.

regroup take apart a number to write it in a new way
1 ten + 2 ones becomes 12 ones.

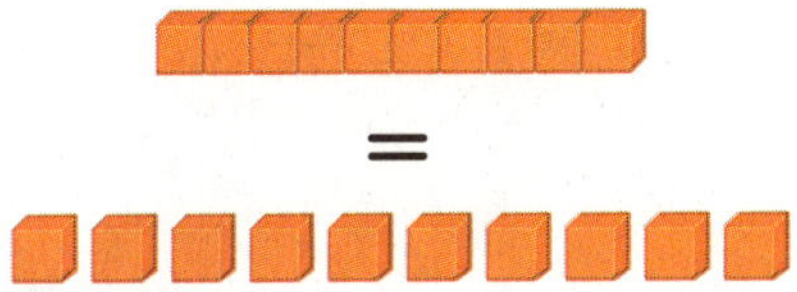

Example 1

Find 75 − 51.

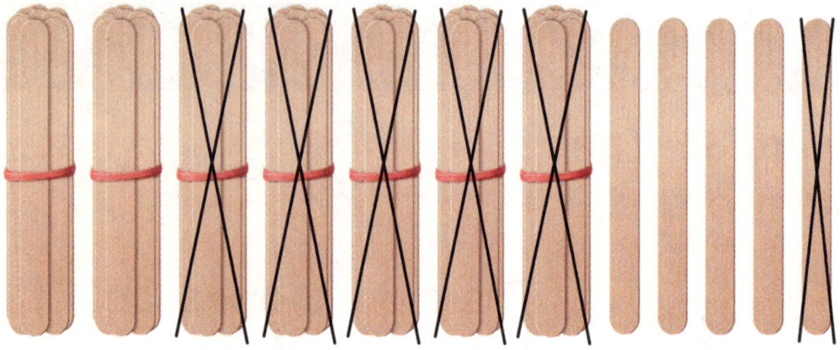

tens	ones
7	5
− 5	1
2	4

Step 1 Subtract the ones column. 5 − 1 = 4
Write 4 in the ones place.

Step 2 Subtract the tens column. 7 − 5 = 2
Write 2 in the tens place.

Answer 75 − 51 = 24

Step-by-Step Practice

Find 64 − 23.

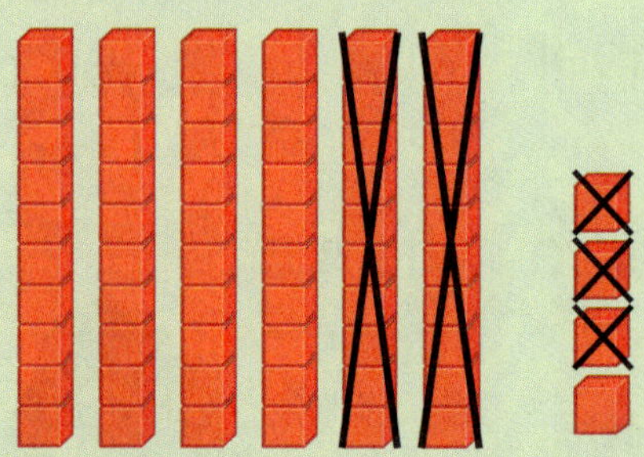

tens	ones
6	4
− 2	3
4	1

Step 1 Subtract the ones column. 4 − 3 = __1__

Write ____ in the ones place.

Step 2 Subtract the tens column. 6 − 2 = ____

Write ____ in the tens place.

Answer ____ − ____ = ____

Example 2

Find 88 − 29.

Step 1 9 is greater than 8. Regroup.

Step 2 Subtract 1 ten from 8 tens. 8 − 1 = 7
Add 10 ones to 8 ones. 10 + 8 = 18

Step 3 Subtract the ones column. 18 − 9 = 9
Write 9 in the ones place.

Step 4 Subtract the tens column. 7 − 2 = 5
Write 5 in the tens place.

Answer 88 − 29 = 59

	tens	ones
	7	18
	~~8~~	~~8~~
−	2	9
	5	9

Step-by-Step Practice

Find 71 − 26.

Step 1 6 is greater than 1. Regroup.

Step 2 Subtract 1 ten from 7 tens. 7 − 1 = ____

Add 10 ones to 1 one. 10 + 1 = ____

Step 3 Subtract the ones column.

____ − ____ = ____

Write ____ in the ones place.

Step 4 Subtract the tens column.

____ − ____ = ____

Write ____ in the tens place.

Answer 71 − 26 = ____

	tens	ones
	☐	☐
	7	1
−	2	6

Name ______________________________

Guided Practice

Find each difference.

1 65 − 22

tens	ones
6	5
− 2	2

2 79 − 14

tens	ones
7	9
−	

3 55 − 31

tens	ones
−	

4 32 − 14

tens	ones
□	□
3	2
− 1	4

5 65 − 29

tens	ones
□	□
6	5
−	

6 84 − 36

tens	ones
□	□
−	

7

□	□
3	7
− 1	8

8

  74
− 33
□

9

□	□
5	1
− 3	6

Problem-Solving Practice

10 There were 33 blocks on the floor. Imena picked up 16 blocks. How many blocks are left on the floor?

Understand Underline key words.

Plan Use a model.

Solve Use base-ten blocks to regroup.
Cross out 16. How many are left?

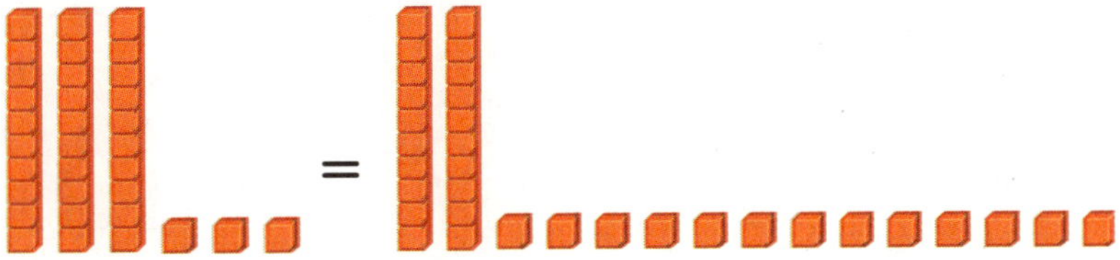

There are _______ blocks left on the floor.

Check Use a hundred chart. Start at 33. Count back 16.

Practice on Your Own

Find each difference.

11 67 − 52

	tens	ones
	6	7
−		

12 75 − 28

	tens	ones
	□	□
	7	5
−		

13 88 − 39

	tens	ones
	□	□
−		

GO on

14 Pam counted 43 birds.
Then 21 birds flew away.
How many birds are left?

There are ______ birds left.

−	

15 Ryan has 21 goldfish in his tank.
He gives 13 of them to his brother.
How many fish does Ryan have left?

Ryan has ______ fish left.

☐	☐
−	

16 WRITING IN ►MATH Explain how to subtract 57 − 18 using a place-value chart. Complete the place-value chart.

	☐	☐
	5	7
−	1	8

____

____

____

____

Vocabulary Check Complete.

17 When you ______________, you take apart a number to write it in a new way.

STOP

Name ________________________________

Progress Check 3 (Lessons 1-5 and 1-6)

Find each sum.

1 34 + 53

	tens	ones
	3	4
+	5	3

2 67 + 18

	tens	ones
	□	
+		

3

	□	
	6	9
+	2	3

Find each difference.

4 79 − 34

	tens	ones
	7	9
−	3	4

5 83 − 29

	tens	ones
	□	□
−		

6

$$\begin{array}{r} 57 \\ -\ 35 \\ \hline \square \end{array}$$

7 There are 19 students in Miss Charnick's third-grade math class. Mr. Cruz has 24 students in his third-grade math class. How many math students are there in all?

_______ math students in all

Name ____________________

«« Replay

M 56 + 13 = ☐

H 73 − 57 = ☐

E 24 + 70 = ☐

A 96 − 45 = ☐

T 67 + 25 = ☐

O 80 − 17 = ☐

C 48 + 34 = ☐

I 49 − 32 = ☐

S 22 + 68 = ☐

Find the sum or difference.
Use the letters to solve the riddle.

What do moths study in school?

Write the letters that match each answer below.

___	___	___	___	___	___	___	___	___	___	___
69	63	92	16	94	69	51	92	17	82	90

Name ______________________________

Review

Vocabulary

Word Bank

- difference
- regroup
- sum

Use the Word Bank to complete.

1. $9 + 1 = 10$ ← ____________

2. $25 - 13 = 12$ ← ____________

3. 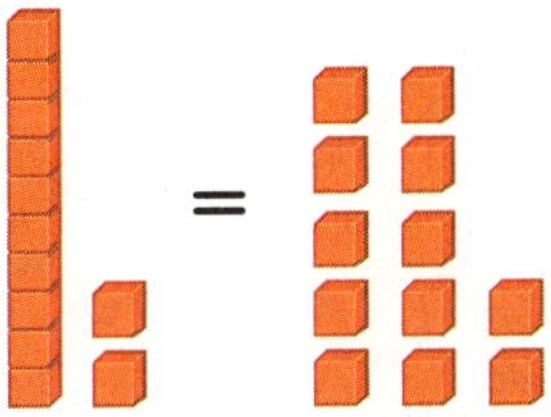 ← ____________

Concepts

Find each sum or difference.

4. 

3 + ____ = ____

5. 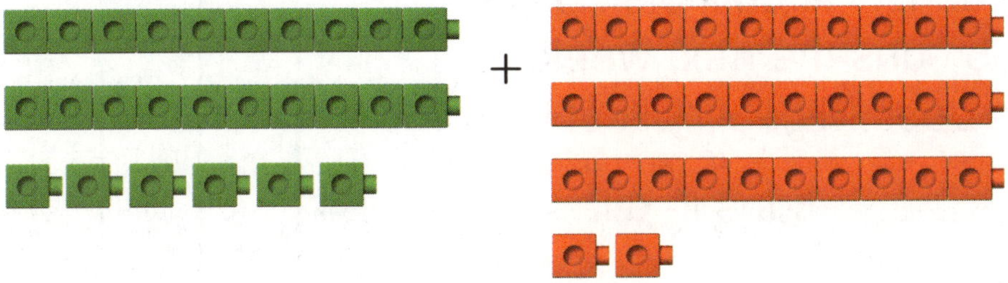

____ + ____ = ____

GO on

Find each sum or difference.

6

____ − ____ = ____

7 49 + 28

	tens	ones
	1	
	4	9
+	2	8
	7	7

8 52 + 34

	5	2
+	3	4

9 56 + 19

```
   □
   56
+  19
-----
```

10 48 − 19

	tens	ones
	□	□
	4	8
−	1	9

11 51 − 33

	□	□
	5	1
−		

12 74 − 26

```
  □ □

−
-----
```

13 Miguel recycles cans. He collects 28 cans one week and 35 cans the next week. How many cans does Miguel collect in all?

Miguel collects ________ cans in all.

	tens	ones
	□	
+		

STOP

Name ______________________________

Chapter Test

Find each sum or difference.

1 $4 + 5 =$ ____

2 $10 - 7 =$ ____

3 $24 + 10 =$ ____

4 $18 - 11 =$ ____

5

	tens	ones
	□	
	5	5
+	2	8

6

	tens	ones
	7	9
−	3	6

7

	tens	ones
	□	□
	8	3
−	3	9

8

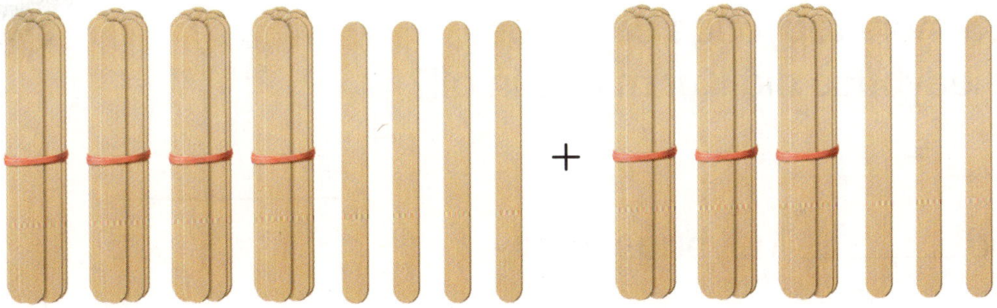

____ + ____ = ____

9

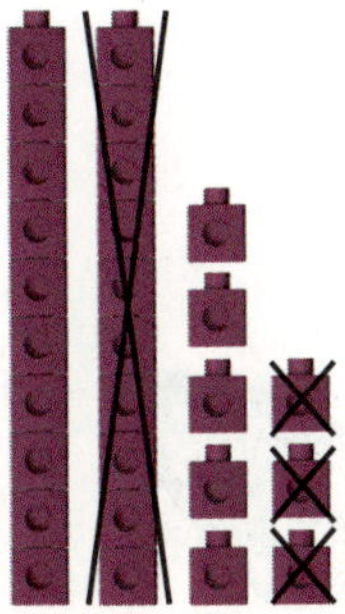

____ − ____ = ____

GO on

10 Who is Correct?

Emma and Jorgé solved the problem 81 − 57 = ____.

Emma

Jorgé

Circle the correct answer. Explain.

____

____

11 Karim found 19 ladybugs in the grass.
He found 23 more ladybugs on a tree.
How many ladybugs did Karim find in all?

____ + ____ = ____

Karim found ______ ladybugs in all.

12 Nicholas had 34 pictures of butterflies.
He gave 26 pictures to his sister.
How many butterfly pictures does Nicholas have left?

____ − ____ = ____

Nicholas has ____ butterfly pictures left.

STOP

Name ______________________

Test Practice

Choose the correct answer.

1 How many ones does 7 tens equal?

○ 0　○ 7　○ 17　○ 70

2 Which of the following is **NOT** true about the addition sentence?

$52 + 17 = 69$

○ The sum is 69.
○ 52 is being added to 17.
○ 17 is an addend.
○ The sum is 52.

3 How many ones does 5 tens equal?

○ 5　○ 15　○ 50　○ 55

4 Which symbol tells you to join sets together?

○ $+$　○ $-$　○ $=$　○ $<$

5 Which number sentence does the number line show?

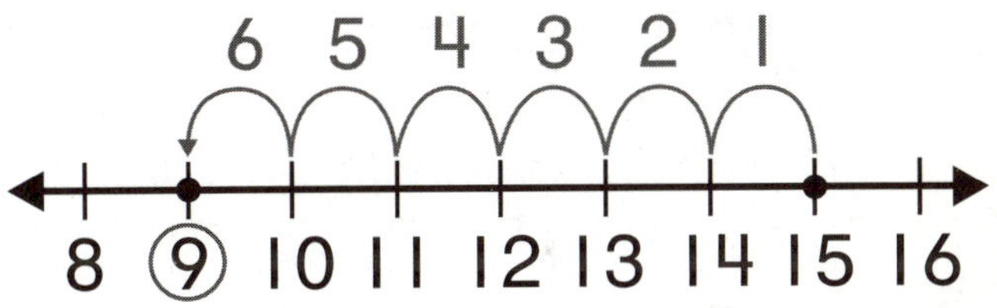

○ $15 - 9 = 6$
○ $15 - 6 = 9$
○ $9 + 6 = 15$
○ $6 + 9 = 15$

6 Which symbol tells you to separate sets?

○ $=$　○ $<$　○ $+$　○ $-$

GO ON

7 Which number sentence is shown by this picture?

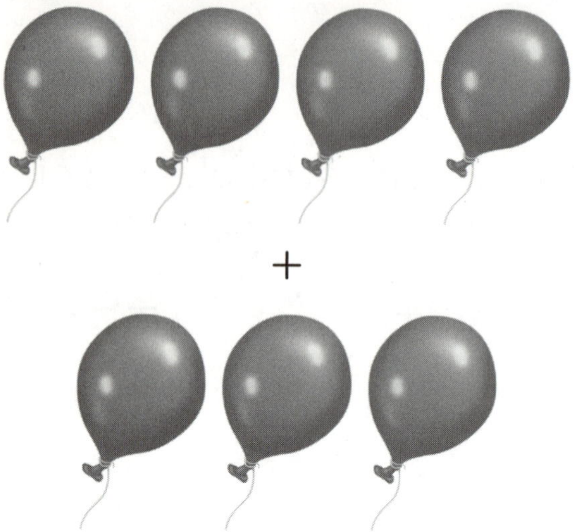

- ○ 4 − 3 = 1
- ○ 7 − 3 = 4
- ○ 4 + 3 = 7
- ○ 3 + 4 = 12

8 What is the sum of 13 and 86?

73	79	96	99
○	○	○	○

9 Complete the subtraction sentence.

57 − 42 = ____

5	15	26	99
○	○	○	○

10 What symbol best compares these amounts?

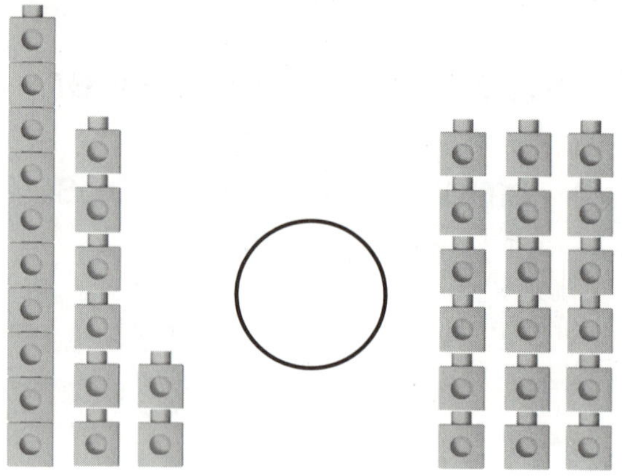

<	>	+	=
○	○	○	○

11 What is the difference of 51 and 38?

13	25	27	89
○	○	○	○

12 Complete the addition sentence.

57 + 19 = ____

38	62	76	616
○	○	○	○

STOP

Introduction to Multiplication

Home Connection

English

Dear Family,

Today our class started **Chapter 2, Introduction to Multiplication.** In this chapter, I will learn about equal groups and skip counting. I will also learn about arrays and area models. I will use these strategies to multiply numbers by 0, 1, 2, and 5.

Love, ________________

Spanish

Estimada Familia:

Hoy en clase empezó **el Capítulo 2, Introducción a la multiplicación.** En este capítulo voy a aprender sobre los grupos iguales y a brincar salteado. También aprenderé sobre los arreglos *(arrays)* y los modelos de área. Usaré estas estrategias para aprender a multiplicar por cero, 1, 2, y 5.

Cariños, ________________

Help at Home

You can practice skip counting with your child at home. Line up pairs of shoes. Skip count by 2s to find the number of shoes in all. Find other objects that can be counted by 2s or 5s, and use skip counting to find how many objects in all.

Math Online Take the chapter Get Ready quiz at macmillanmh.com.

Ayude en casa

Usted puede practicar contar salteado con su niño en casa. Puede alinear pares de zapatos. Cuente salteado, de 2 en 2, para calcular el número total de zapatos. Busque otros objetos que puedan ser contados de 2 en 2, de 5 en 5, y cuente salteado para calcular el total de objetos.

Name ______________________________

Get Ready

Count. Write how many.

1 There are _____ rows of buttons.

There are _____ buttons in each row.

There are _____ buttons in all.

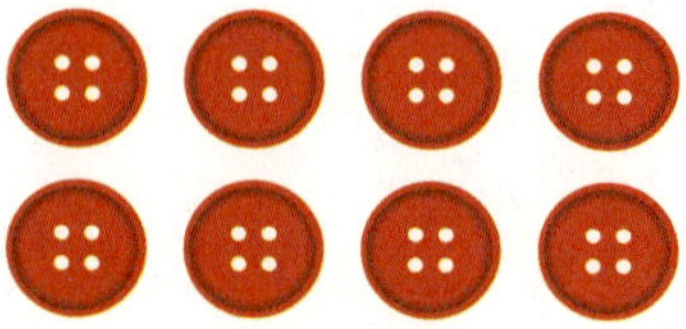

2 There are _____ rows of stickers.

There are _____ stickers in each row.

There are _____ stickers in all.

3 The rectangle has _____ rows.

Each row has _____ squares.

There are _____ squares in all.

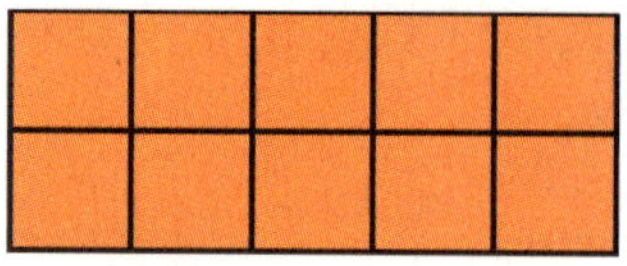

4 The square has _____ rows.

Each row has _____ small squares.

There are _____ small squares in all.

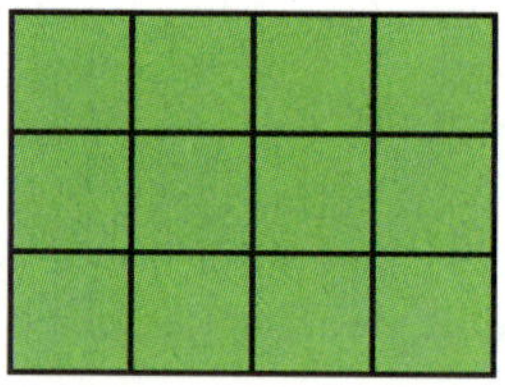

Find each sum.

5 $2 + 2 + 2 =$ _____

6 $4 + 4 =$ _____

7 $3 + 3 =$ _____

8 $5 + 5 + 5 =$ _____

STOP

Lesson 2-1

Name ____________________

Equal Groups

Key Concept

Alicia has 8 buttons.

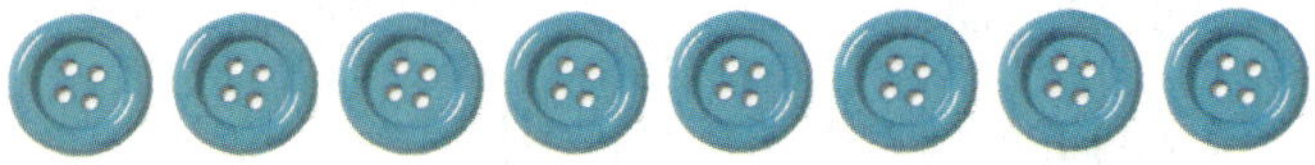

Alicia can place the buttons in 2 **equal groups**. There are 4 buttons in each group.

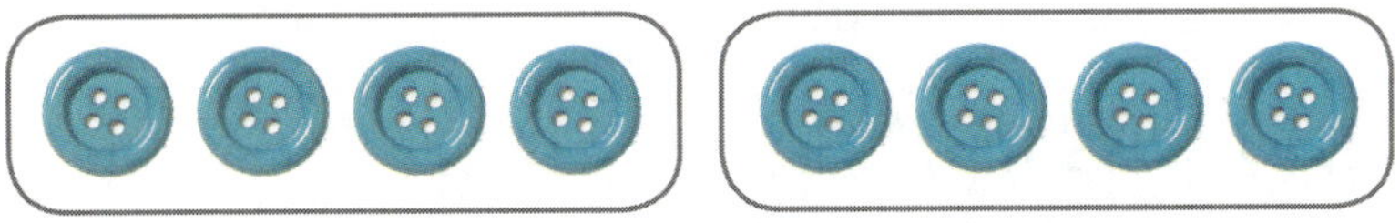

4 4

2 groups of 4 = 8 buttons in all

Vocabulary

equal groups each group has the same number of objects

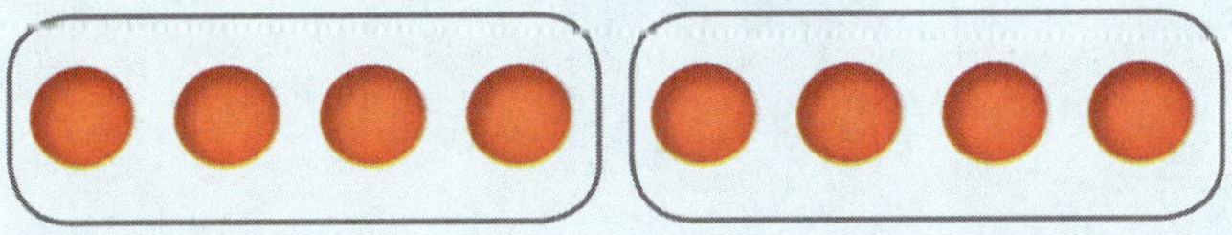

There are 2 equal groups of counters.

There are 4 counters in each group. There are 8 counters in all.

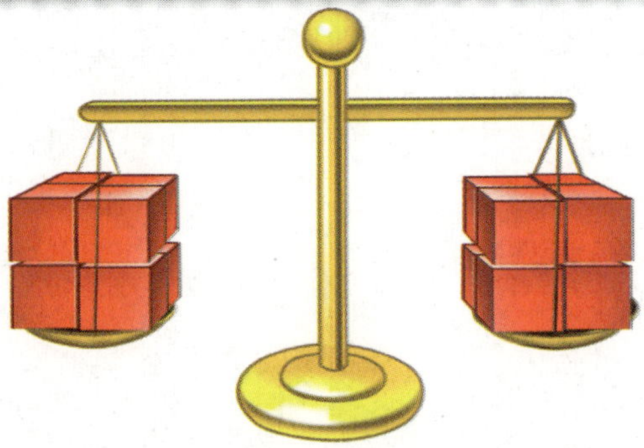

There are 4 cubes on each side. There are 8 cubes in all.

Example

Write how many in all.

3 3 3

3 groups of 3 9 cowboy hats in all

Step 1 Count the number of groups. 3

Step 2 Count how many in each group. 3

Step 3 Find the total. 9

Answer There are 9 cowboy hats in all.

Step-by-Step Practice

Write how many in all.

_______ _______

_____ groups of _____ _____ cowboy boots in all

Step 1 Count the number of groups. _____

Step 2 Count how many in each group. _____

Step 3 Find the total. _____

Answer There are _____ cowboy boots in all.

Name ______________________________

Guided Practice

Write how many in all.

1 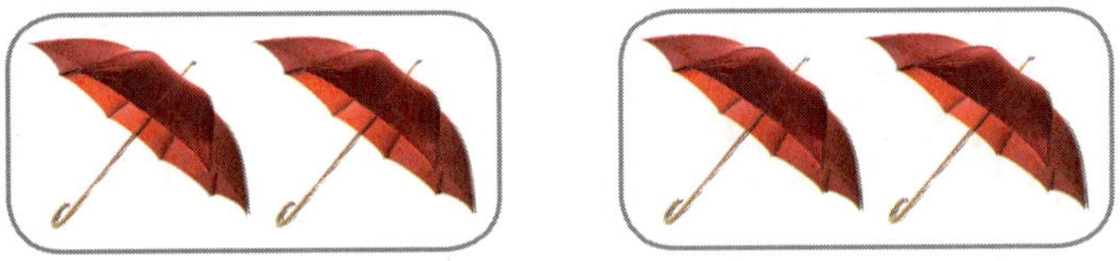

_____ groups of _____ _____ umbrellas in all

2 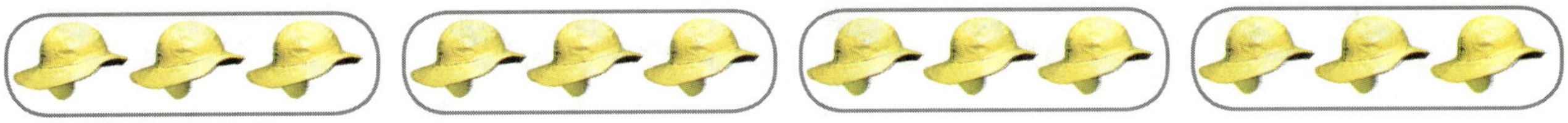

_____ groups of _____ _____ rain hats in all

Problem-Solving Practice

3 Vincent watches movies on a rainy day. He has 4 types of movies. He has 4 movies of each type. How many movies does Vincent have in all?

Understand Underline key words.

Plan Draw a picture.

Solve Draw _____ groups of _____ rectangles.

Vincent has _______ movies in all.

Check Circle each group of rectangles. Skip count by 4s to check your answer.

GO on

Practice on Your Own

Write how many in all.

4 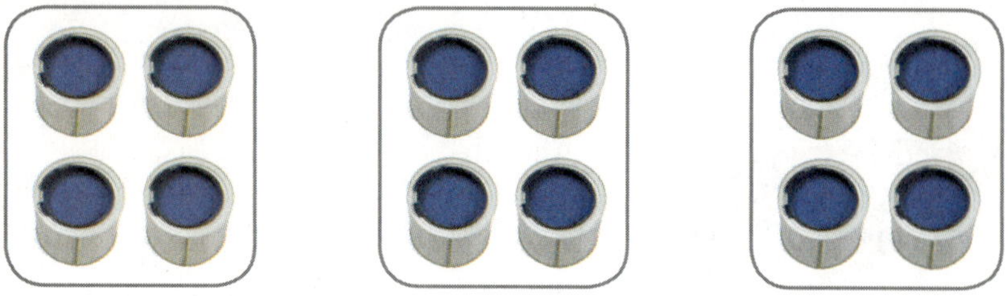

_____ groups of _____ _____ paint cans in all

5 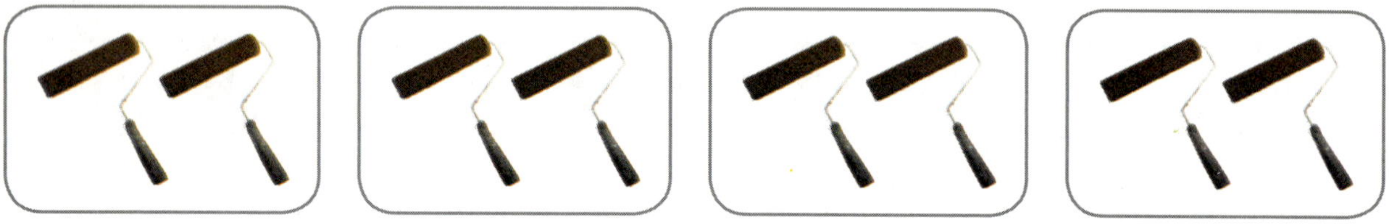

_____ groups of _____ _____ paint rollers in all

6 **WRITING IN MATH** Elaine has 10 paintbrushes. Can Elaine place the paintbrushes in equal groups? How do you know?

____

____

Vocabulary Check Complete.

7 When each group has the same number of objects, they are _____________________.

STOP

Lesson 2-2

Name ______________________________

Repeated Addition and Skip Counting

Key Concept

You can use **repeated addition** to find the total.

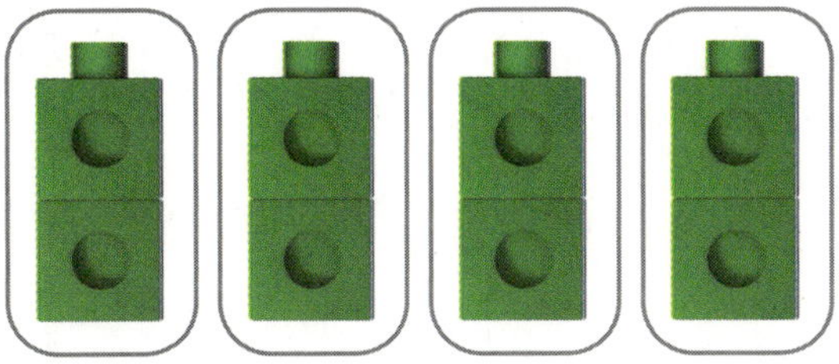

2 + 2 + 2 + 2 = 8

You can also **skip count** to find the total.

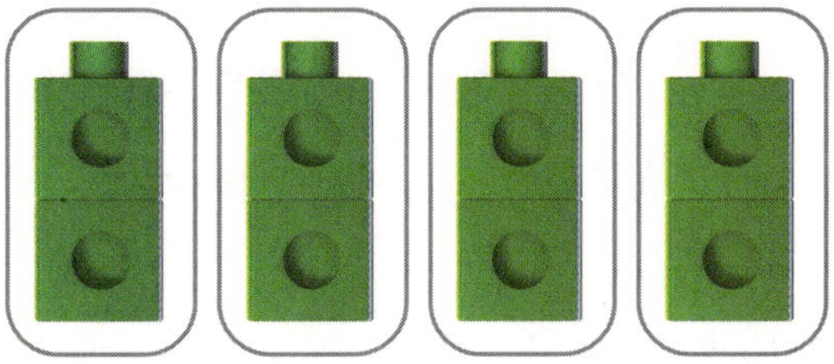

2, 4, 6, 8

There are 8 cubes in all.

Vocabulary

repeated addition add the same number more than one time

skip count count objects in equal groups of two or more 2, 4, 6, 8, 10

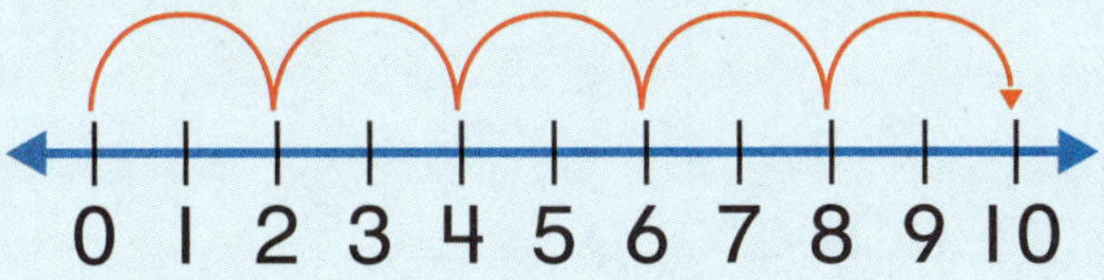

Example

Skip count to find the sum.

Step 1 Count the number in each group. 2

Step 2 Count the number of groups. 5

Step 3 Write an addition sentence.

2 + 2 + 2 + 2 + 2 = 10

Step 4 Skip count to find the sum. 2, 4, 6, 8, 10

Answer There are 10 frogs in all.

Step-by-Step Practice

Skip count to find the sum.

Step 1 Count the number in each group. ____

Step 2 Count the number of groups. ____

Step 3 Write an addition sentence.

____ + ____ + ____ = ____

Step 4 Skip count to find the sum. ____, ____, ____

Answer There are ____ bees in all.

Name ____________________

Guided Practice

1 Use repeated addition and skip counting to find the sum.

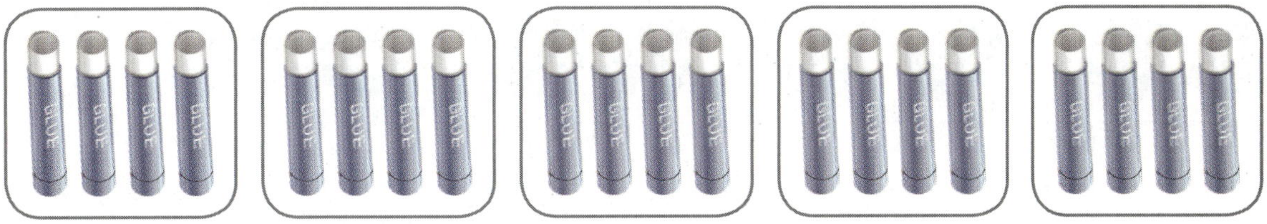

____ + ____ + ____ + ____ + ____ = ____

4, ____, ____, ____, ____

There are ____ glue sticks in all.

Problem-Solving Practice

2 Alfonso has 5 boxes of crayons. There are 5 crayons in each box. How many crayons does Alfonso have in all?

Understand Underline key words.

Plan Write an addition sentence.

Solve Find the sum.

____ + ____ + ____ + ____ + ____ = ____

Alfonso has ____ crayons in all.

Check Skip count by 5s.

Practice on Your Own

Use repeated addition and skip counting to find each sum.

3

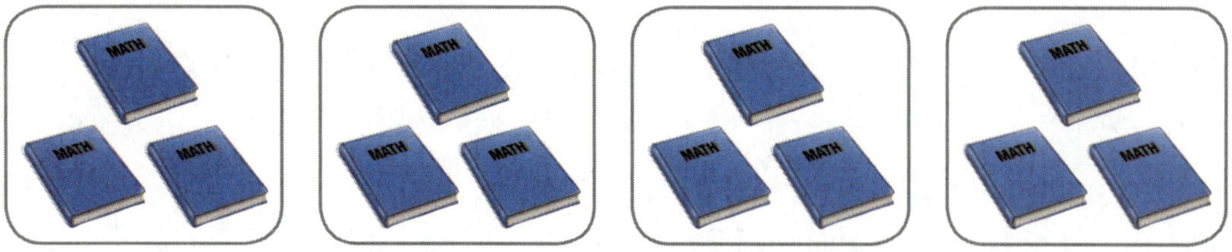

____ + ____ + ____ + ____ = ____

3, ____, ____, ____

There are ____ math books in all.

4

____ + ____ + ____ = ____

5, ____, ____

There are ____ reading books in all.

5 WRITING IN MATH Domingo put 10 books on each of 3 shelves. How many books did he put on the shelves in all? Explain.

____

____

____

Vocabulary Check Complete.

6 You can ______________ to count objects in equal groups of two or more.

STOP

Name ______________________________

Progress Check 1 (Lessons 2-1 and 2-2)

1 Write how many in all.

____ groups of ____ There are ____ deer in all.

2 Use repeated addition and skip counting to find the sum.

____ + ____ + ____ + ____ + ____ = ____

____, ____, ____, ____, ____

There are ____ bears in all.

3 There are 6 exhibits of monkeys at a zoo. There are 3 monkeys in each exhibit. How many monkeys are on exhibit in all? Explain.

____

____

____

____

Name ______________________________

«« Replay Code Combination

Use skip counting and repeated addition to find the secret code and unlock the gym locker.

Kofi's football team made 5 field goals.
A field goal is worth 3 points.

Skip count. ____, ____, ____, ____, ____ points

The team scored ☐ points from field goals.

The team made 3 touchdowns. A touchdown is worth 6 points.

Use repeated addition. ____ + ____ + ____ = ____ points

The team scored ☐ points from touchdowns.

How many points did the team score in all? Add.

____ + ____ = ____

The team scored ☐ points from touchdowns.

Lesson 2-3

Name ______________________________

Arrays

Key Concept

You can use an **array** to show multiplication.

There are 2 rows.
There are 4 in each row.
The **product** is 8.

2	×	4	=	8
number of rows		number in each row		in all

Vocabulary

array objects shown in rows and columns

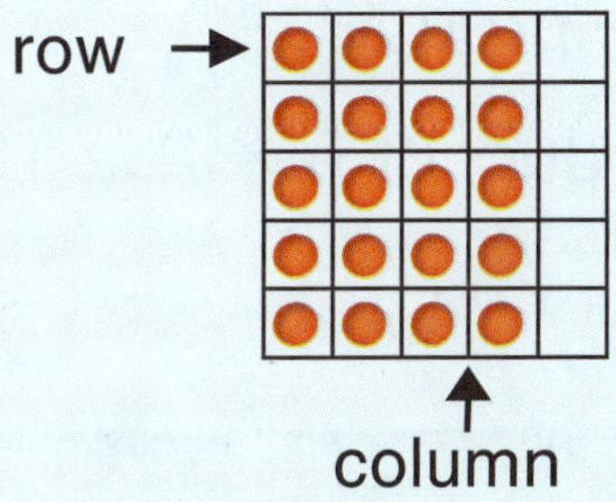

product the answer to a multiplication problem

$3 \times 4 = 12$ ↑

multiplication sentence a math sentence that has a multiplication sign in it

$5 \times 4 = 20$

You can check multiplication with repeated addition.

$4 \times 6 = 24$

$4 + 4 + 4 + 4 + 4 + 4 = 24$

Example

Write a multiplication sentence. Find the product.

5	×	2	=	10
number of rows		number in each row		in all

Step 1 Count the number of rows. 5

Step 2 Count the number in each row. 2

Step 3 Find the product. 10

Answer The product of 5 × 2 is 10.

Step-by-Step Practice

Write a multiplication sentence. Find the product.

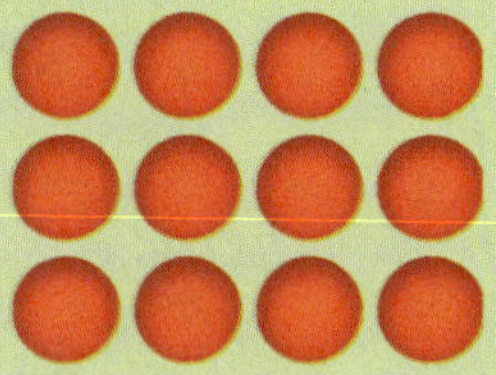

3	×	4	=	12
number of rows		number in each row		in all

Step 1 Count the number of rows. ____

Step 2 Count the number in each row. ____

Step 3 Find the product. ____

Answer The product of ____ × ____ is ____.

Name ______________________________

Guided Practice

Write a multiplication sentence. Find each product.

1 

____ × ____ = ____

rows, in each row, in all

2 

____ × ____ = ____

rows, in each row, in all

Problem-Solving Practice

3 There are 3 picnic baskets.
There are 6 bananas in each basket.
How many bananas are there in all?

Understand Underline key words.

Plan Draw an array.

Solve

Write a multiplication sentence.

____ × ____ = ____

There are ____ bananas in all.

Check Use repeated addition to check your answer.

GO on

Practice on Your Own

Write a multiplication sentence. Find each product.

4 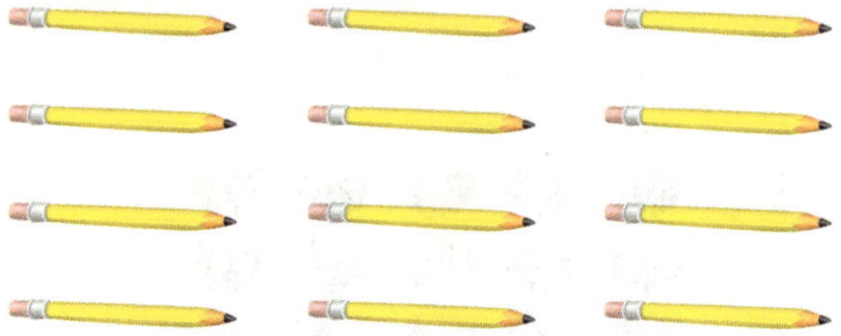

_____ × _____ = _____

5 

_____ × _____ = _____

6 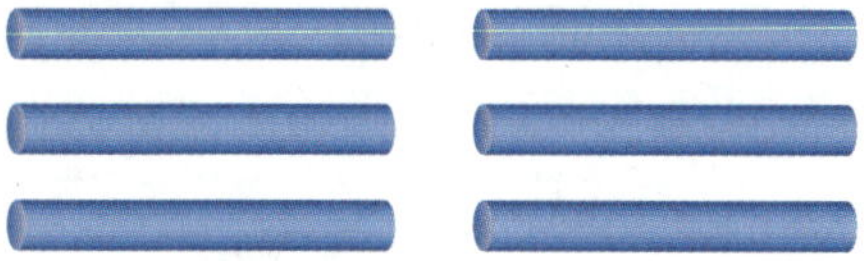

_____ × _____ = _____

7 

_____ × _____ = _____

8 **WRITING IN MATH** There are 4 rows of desks in Mr. Oliver's classroom. There are 6 desks in each row. Molly says there are 10 desks in all. Is Molly correct? How do you know?

____

____

____

Vocabulary Check Complete.

9 In 2 × 3 = 6, 6 is the _______________.

10 An _______________ shows objects in rows and columns.

STOP

Lesson 2-4

Name ______________________________

Area Models

Key Concept

You can use an **area model** to show multiplication.

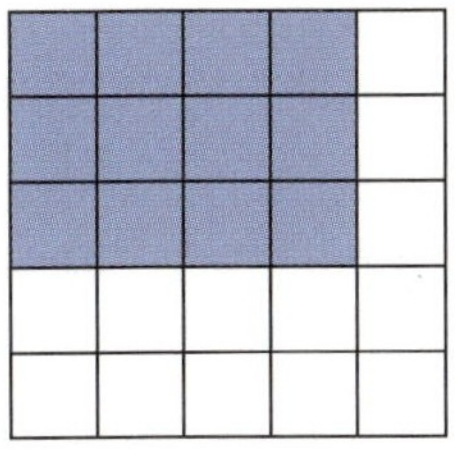

3	×	4	=	12
number of rows		number in each row		in all

There are 3 shaded rows.
There are 4 shaded squares in each row.
There are 12 shaded squares in all.
The product is 12.

Vocabulary

area model a grid that models a multiplication sentence

$3 \times 5 = 15$

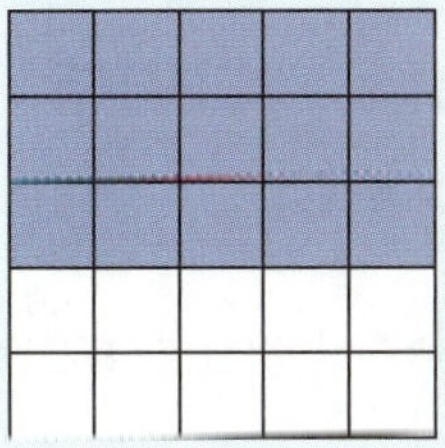

factor a number that is multiplied by another number

$3 \times 6 = 18$

↑ ↑

Peyton collects state quarters.
She has filled 3 rows of quarters, with 4 quarters in each row.
Peyton has collected 12 state quarters in all.

Example

Shade the area model.
Find the product.

4 × 2

Step 1 The first factor is 4.

Step 2 The second factor is 2.

Step 3 Shade 4 rows.
Shade 2 squares in each row.

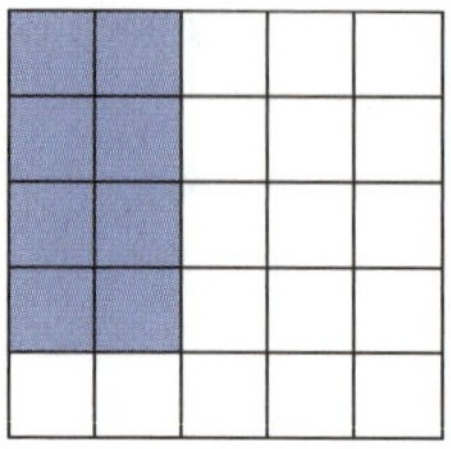

Step 4 Count the shaded squares to find the product.

Answer $4 \times 2 = 8$

Step-by-Step Practice

Shade the area model.
Find the product.

2 × 3

Step 1 The first factor is ____.

Step 2 The second factor is ____.

Step 3 Shade ____ rows.

Shade ____ squares in each row.

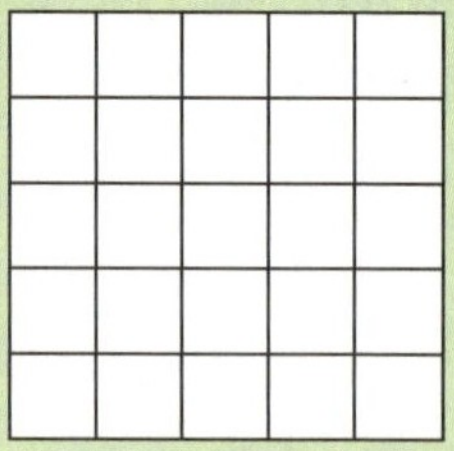

Step 4 Count the shaded squares to find the product.

Answer $2 \times 3 =$ ____

Name ______________________________

Guided Practice

Shade the area model. Find each product.

1 3 rows of 5

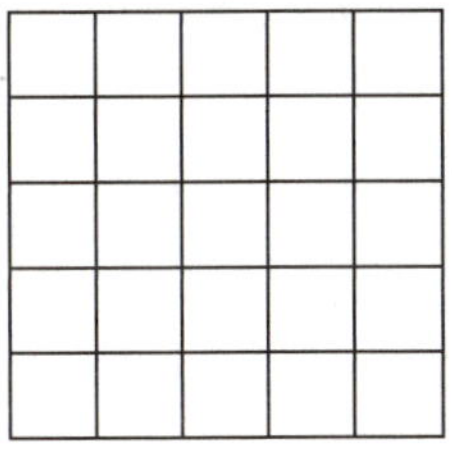

$3 \times 5 =$ ____

2 5 rows of 2

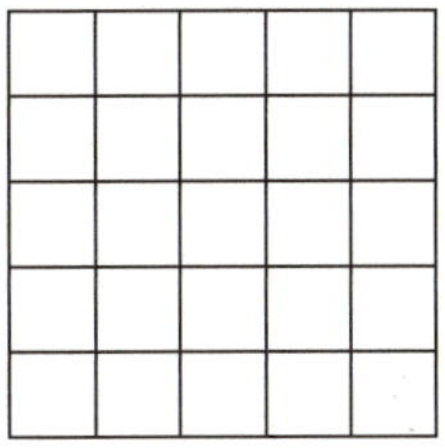

$5 \times 2 =$ ____

Problem-Solving Practice

3 Shawn fills 5 glasses with ice cubes.
He puts 5 ice cubes in each glass.
How many ice cubes does Shawn use in all?

Understand Underline key words.

Plan Use an area model.

Solve

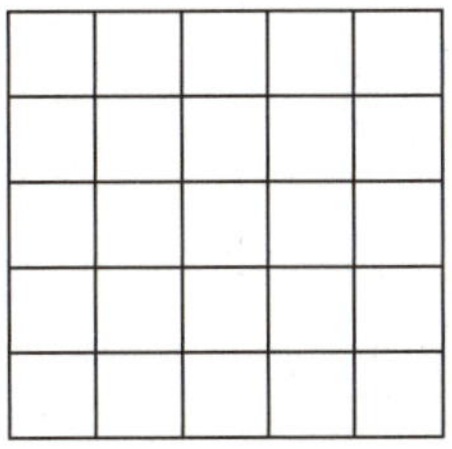

____ glasses × ____ ice cubes in each = ____

Shawn uses ____ ice cubes in all.

Check Skip count by 5s.

GO on

Practice on Your Own

Shade the area model. Find each product.

4 ____ rows of ____

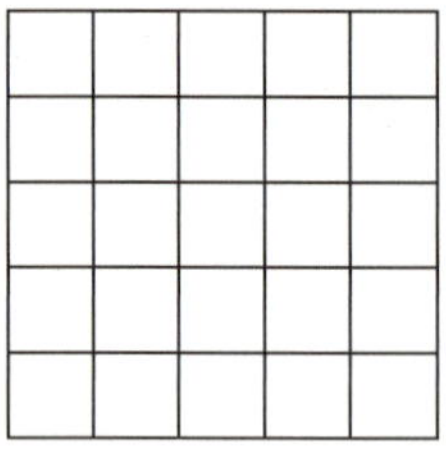

$2 \times 4 =$ ____

5 ____ rows of ____

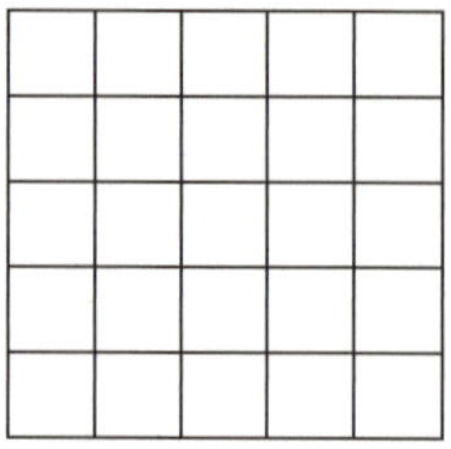

$2 \times 2 =$ ____

6 ____ rows of ____

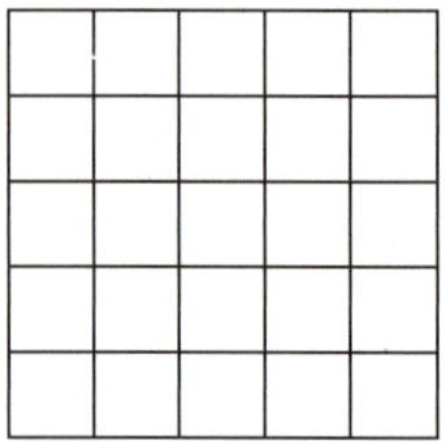

$4 \times 4 =$ ____

7 ____ rows of ____

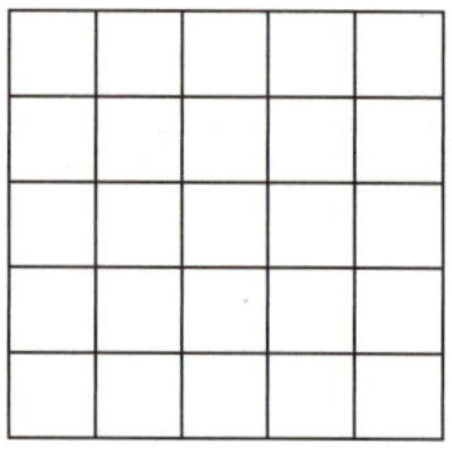

$5 \times 3 =$ ____

8 **WRITING IN MATH** Look at the area model. Write a multiplication sentence. Explain.

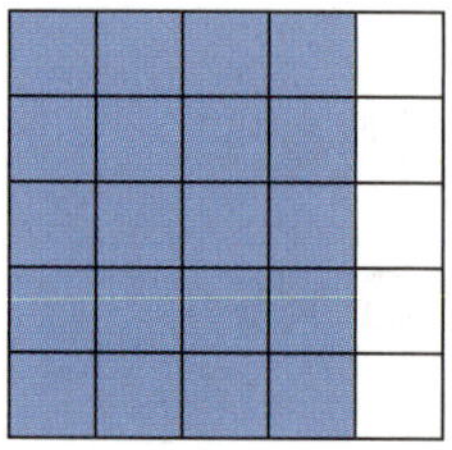

____

____

____

____

Vocabulary Check Complete.

9 This ____________________ shows $3 \times 2 = 6$.

STOP

Name ______________________________

Progress Check 2 (Lessons 2-3 and 2-4)

Write a multiplication sentence. Find each product.

1

_____ × _____ = _____

2

_____ × _____ = _____

Shade the area model. Find each product.

3 _____ rows of _____

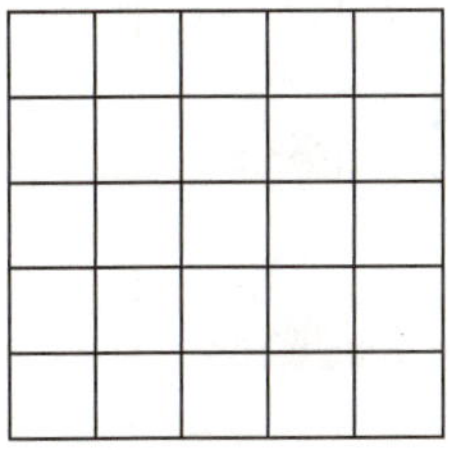

4 × 1 = _____

4 _____ rows _____

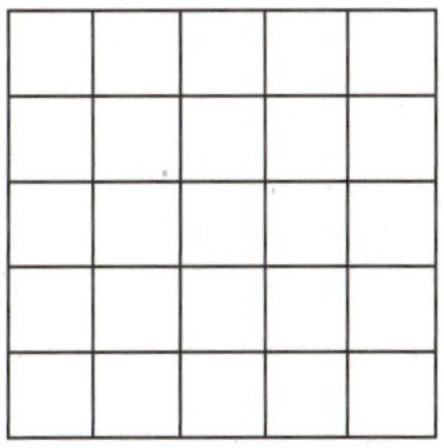

3 × 3 = _____

5 Miss Sung creates a bulletin board. She makes 4 rows of photos. She places 5 photos in each row. How many photos does Miss Sung use in all? How do you know?

____

____

6 Marisol models 4 × 3 = 12. Does he shade the area model correctly? Explain.

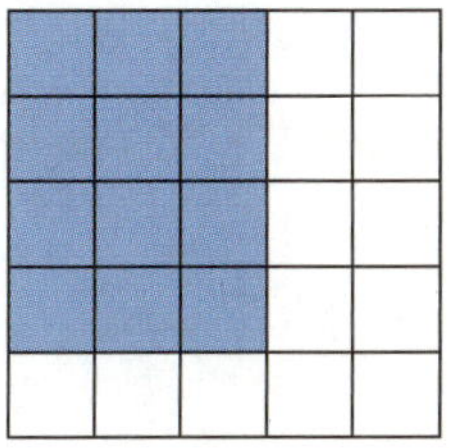

____

____

____

Name ________________________________

Replay Array Match

Draw a line to match each number sentence with the correct array. Find each product.

$2 \times 3 =$ ____

$3 \times 6 =$ ____

$5 \times 1 =$ ____

$5 \times 4 =$ ____

$1 \times 7 =$ ____

Lesson 2-5

Name ______________________________

Multiply by 0 and 1

Key Concept

When you **multiply** any number by 1, the **product** is that number.

$3 \times 1 = 3$ The product of 3×1 is 3.

When you multiply any number by 0, the product is 0.

$3 \times 0 = 0$ The product of 3×0 is 0.

Vocabulary

multiply (multiplication) the operation of repeated addition of the same number

product the answer to a multiplication problem $3 \times 4 = 12$ ↑ (12)

factor a number that is multiplied by another number $3 \times 6 = 18$ ↑ (3) ↑ (6)

When you multiply, you join equal groups.

5	×	1	=	5
number of vases		number of flowers in each vase		product

Example

Find the product.

6 × 1 = 6

baskets — apple in each basket — product

Step 1 The first factor tells how many groups. 6

Step 2 The second factor tells how many in each group. 1

Step 3 Multiply. 6 × 1 = 6

Answer There are 6 apples in all.

Step-by-Step Practice

Find the product.

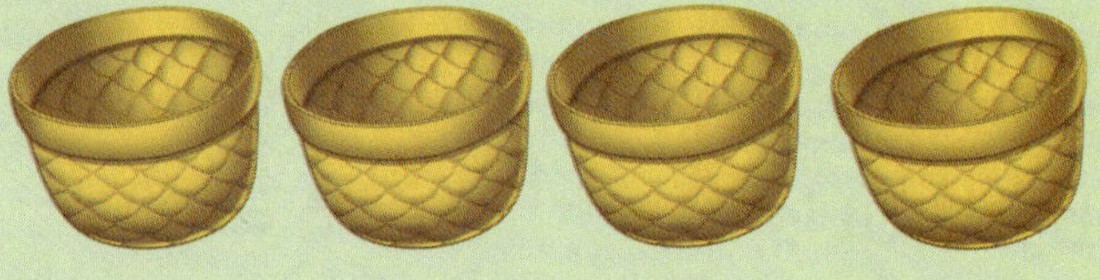

____ × ____ = ____

baskets — apples in each basket — product

Step 1 The first factor tells how many groups. ____

Step 2 The second factor tells how many in each group. ____

Step 3 Multiply. ____ × ____ = ____

Answer There are ____ apples in all.

Name ______________________________

Guided Practice

Find each product.

1 

2 × 1 = ____

number of groups — number in each group — product

2 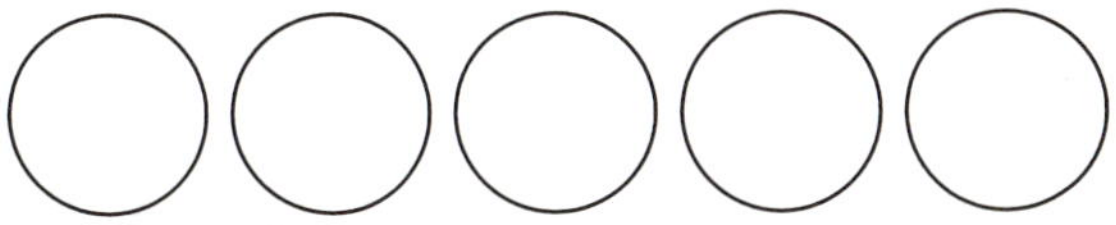

5 × 0 = ____

number of groups — number in each group — product

3 7 × 1 = ____

4 0 × 1 = ____

Problem-Solving Practice

5 Keshia invites 6 friends to her birthday party. She has 1 balloon to give each friend. How many balloons does Keshia have?

Understand Underline key words.

Plan Draw an array.

Solve Write a multiplication sentence.

____ × ____ = ____

Keshia has ____ balloons.

Check Use repeated addition to check your answer.

GO on

Practice on Your Own

Find each product.

6 

____ × ____ = ____

number of groups | number in each group | product

7 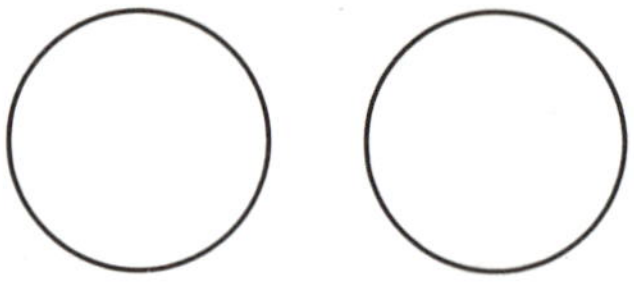

____ × ____ = ____

number of groups | number in each group | product

8 $1 \times 1 =$ ____

9 $0 \times 7 =$ ____

10 $9 \times 1 =$ ____

11 $10 \times 0 =$ ____

12 $1 \times 10 =$ ____

13 $1 \times 3 =$ ____

14 **WRITING IN MATH** Write a multiplication sentence about the dogs. Explain.

____

____

Vocabulary Check Complete.

15 Look at the multiplication sentence $0 \times 9 = 0$.

The number 9 is a ____________.

STOP

Lesson 2-6

Name ______________________________

Multiply by 2 and 5

Key Concept

You can use skip counting to **multiply**.

Find the **product** of 4 and 2.
Skip count by 2s four times.

2, 4, 6, 8

$4 \times 2 = 8$

The product is 8.

Find the product of 3 and 5.
Skip count by 5s three times.

5, 10, 15

$3 \times 5 = 15$

The product is 15.

Vocabulary

multiply (multiplication) the operation of repeated addition of the same number

product the answer to a multiplication problem

$3 \times 4 = 12$ ↑

I can use a number line to skip count by 2s four times.
$4 \times 2 = 8$

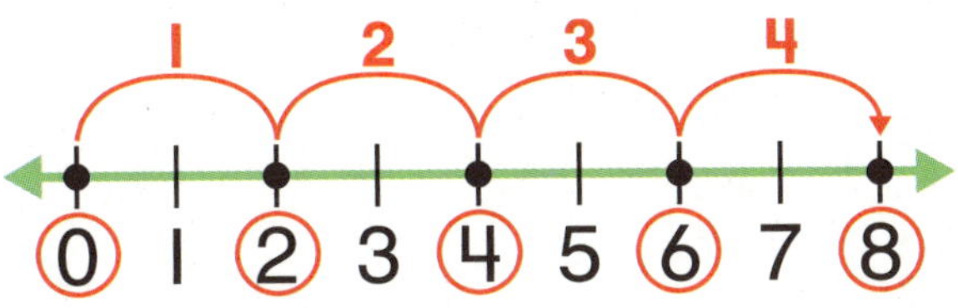

Example

Skip count to find the product. $2 \times 7 =$ ____

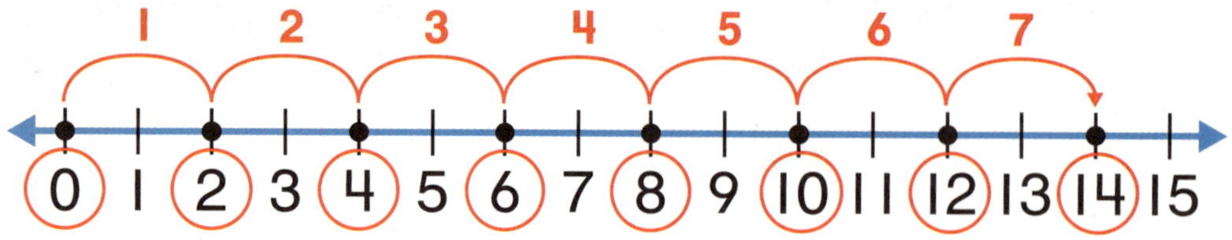

Step 1 Skip count by 2s.

Step 2 Skip count 7 times.
2, 4, 6, 8, 10, 12, 14

Step 3 Multiply. $2 \times 7 = 14$

Answer The product is 14.

Step-by-Step Practice

Skip count to find the product. $6 \times 5 =$ ____

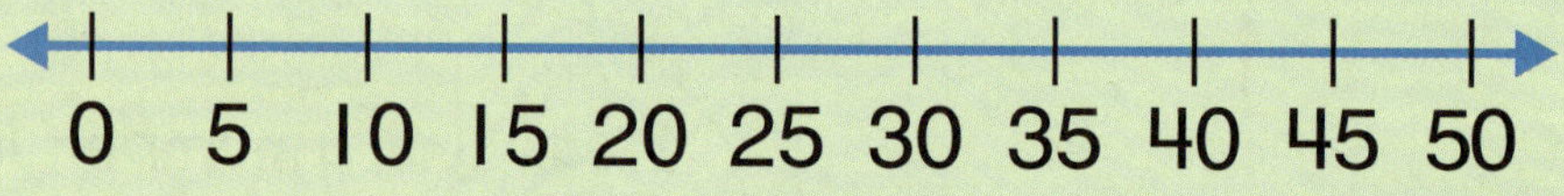

Step 1 Skip count by ____s.

Step 2 Skip count ____ times.
5, ____, ____, ____, ____, ____

Step 3 Multiply. $6 \times 5 =$ ____

Answer The product is ____.

Name ______________________________

Guided Practice

Skip count to find each product. Use the number line.

1 2 × 2 = ____ ____, ____

0 1 2 3 4 5 6 7 8 9 10

2 4 × 5 = ____ ____, ____, ____, ____

0 1 2 3 4 5 6 7 8 9 10 11 12 13 14 15 16 17 18 19 20

Problem-Solving Practice

3 Bruce has 8 pairs of socks.
How many socks does Bruce have in all?

Understand Underline key words.

Plan Skip count.

Solve Skip count by ____s eight times.

____, ____, ____, ____, ____, ____, ____, ____

____ × ____ = ____ socks in all

Check Draw a picture. Count the socks.

GO on

Practice on Your Own

Skip count to find each product.

4 $2 \times 6 =$ ____ ____, ____, ____, ____, ____, ____

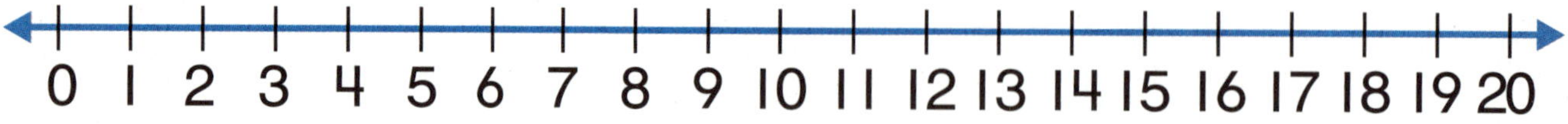

5 $5 \times 5 =$ ____ ____, ____, ____, ____, ____

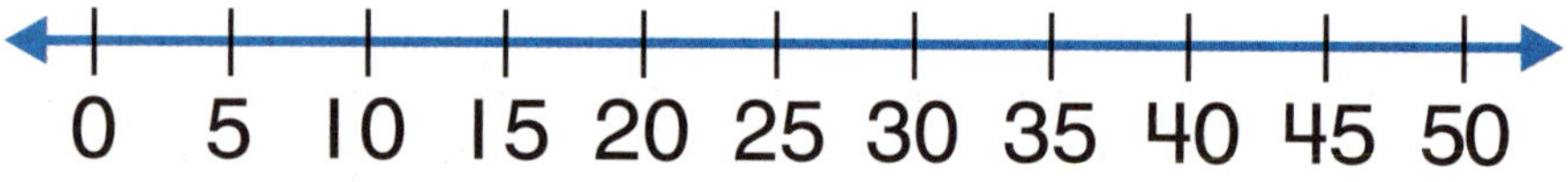

6 $2 \times 4 =$ ____

____, ____, ____, ____

7 $5 \times 3 =$ ____

____, ____, ____

8 WRITING IN MATH Joel has 10 boxes of books. There are 5 books in each box. How many books does Joel have? Explain.

____

____

Vocabulary Check Complete.

9 You ______________ to find a product.

STOP

Name ______________________________

Progress Check 3 (Lessons 2-5 and 2-6)

Find each product.

1 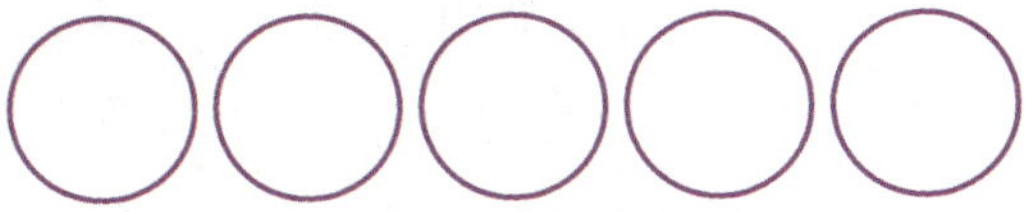

5 × 0 = ____

number of groups · number in each group · product

2 3 × 1 = ____

3 0 × 1 = ____

Skip count to find each product.

4 5 × 4 = ____ ____, ____, ____, ____

5 2 × 2 = ____ ____, ____

6 6 × 5 = ____ ____, ____, ____, ____, ____, ____

7 The problem 2 × 5 appears on a test.
Nashoba skip counts by 2s to find the product. 2, 4, 6, 8, 10.
Mona skip counts by 5s to find the product. 5, 10.
Who is correct? Explain.

____

____

____

Name ____________________

«« Replay Who am I?

Find the vocabulary term or number that answers each riddle.

Answer Bank		
factor	1	9
multiply	4	10
product	0	7
skip count	5	12

1. I am the answer to a multiplication problem. Who am I? __________

2. When any number is multiplied by 0, I am the product. Who am I? __________

3. When I am multiplied by another number, the product is always that number. Who am I? __________

4. I am a number that is multiplied by another number. Who am I? __________

5. I am the product of 5×2. Who am I? __________

6. I am the product of 1×4. Who am I? __________

7. I am the product of 9×1. Who am I? __________

Name ______________________________

Review

Vocabulary

Word Bank

array

factor

product

Use the Word Bank to complete.

1  ← ____________

2 $4 \times 2 = 8$ ← ____________

3 $3 \times 5 = 15$ ↑ ____________

Concepts

4 Write how many in all.

____ groups of ____ ____ chicks in all

5 Use repeated addition and skip counting to find the sum.

____ + ____ + ____ = ____ penguins

____, ____, ____ penguins

GO on

6 Write a multiplication sentence. Find the product.

____ × ____ = ____

7 Shade the area model. Find the product.

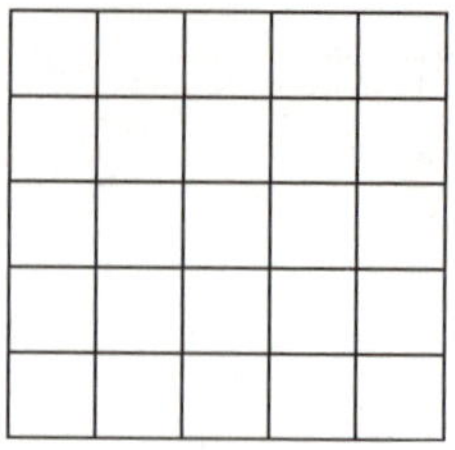

$4 \times 4 =$ ____

Find each product.

8 $1 \times 4 =$ ____

9 $0 \times 8 =$ ____

10 $1 \times 0 =$ ____

Skip count to find each product.

11 $2 \times 6 =$ ____ ____, ____, ____, ____, ____, ____

12 $5 \times 4 =$ ____ ____, ____, ____, ____

13 Leslie sees 5 groups of birds. There are 5 birds in each group. How many birds does Leslie in all? Explain.

____

____

____

STOP

Name ______________________

Chapter Test

Write how many in all.

1

____ groups of ____

____ cubes in all

2

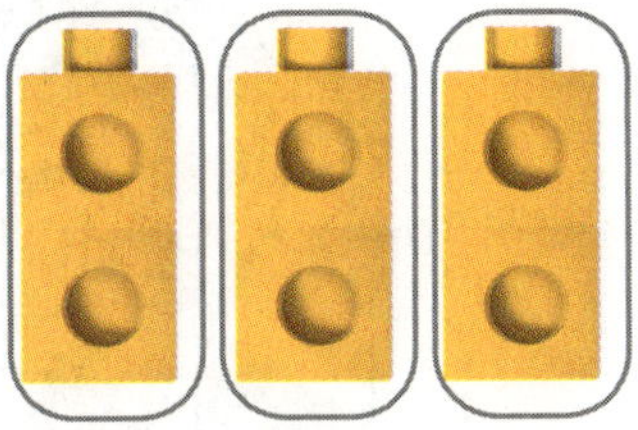

____ groups of ____

____ cubes in all

3 Use repeated addition and skip counting to find the sum.

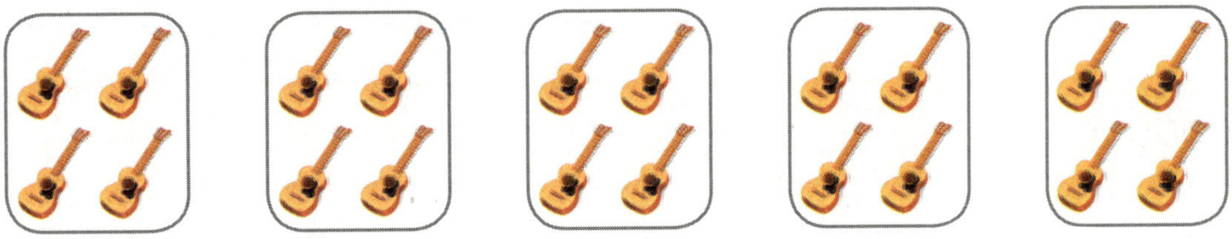

____ + ____ + ____ + ____ + ____ = ____

____, ____, ____, ____, ____

Write a multiplication sentence. Find each product.

4

____ × ____ = ____

5

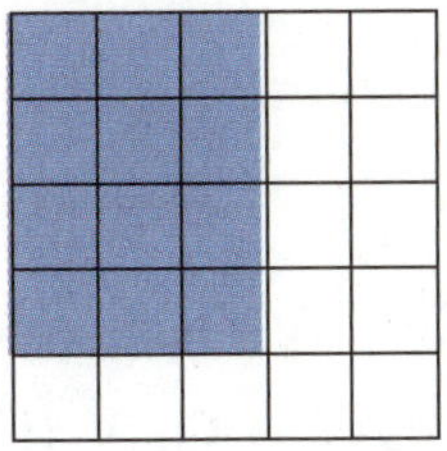

____ × ____ = ____

6 Shade the area model.
Find the product.

5 × 3 = ____

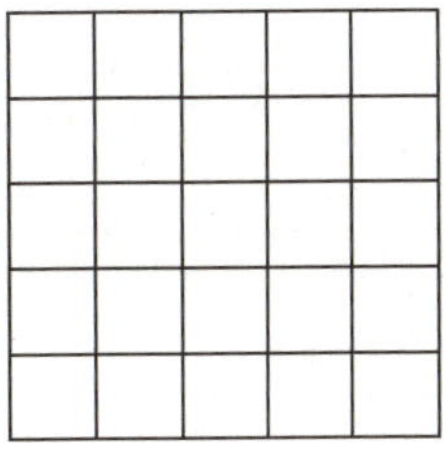

GO on

7 Who is Correct?

Rodolfo and Desta find the product of 0 and 5.

Rodolfo

Desta

Circle the correct answer. Explain.

____

____

Find each product.

8 $1 \times 10 =$ ____ **9** $3 \times 0 =$ ____ **10** $8 \times 1 =$ ____

Skip count to find each product.

11 $4 \times 2 =$ ____ ____, ____, ____, ____

12 $5 \times 5 =$ ____ ____, ____, ____, ____, ____

13 The Smith family has 2 fish tanks. There are 5 fish in each tank. How many fish does the Smith family have in all? How do you know?

____

____

STOP

Name ______________________________

Test Practice

Choose the correct answer.

1 What is the product of 6×1?

○ 0 ○ 1 ○ 6 ○ 7

2 Miss Miller packs for a 5-day trip. She packs 2 shirts for each day. How many shirts does Miss Miller pack in all?

○ 2 ○ 5 ○ 7 ○ 10

3 Which equal groups are modeled?

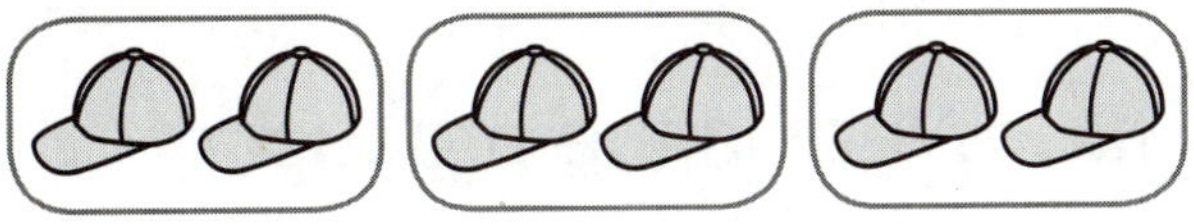

○ 3 groups of 2
○ 4 groups of 2
○ 2 groups of 3
○ 3 groups of 3

4 Which multiplication sentence is true?

○ $2 \times 3 = 5$
○ $1 \times 5 = 5$
○ $0 \times 5 = 5$
○ $1 \times 4 = 5$

5 What is the product of 4×0?

○ 0 ○ 1 ○ 4 ○ 10

6 Emilio has 2 twin sisters. For their birthday, he wants to buy each sister 2 presents. How many presents should Emilio buy in all?

○ 2 ○ 4 ○ 6 ○ 8

GO ON

7 Which multiplication sentence does the area model show?

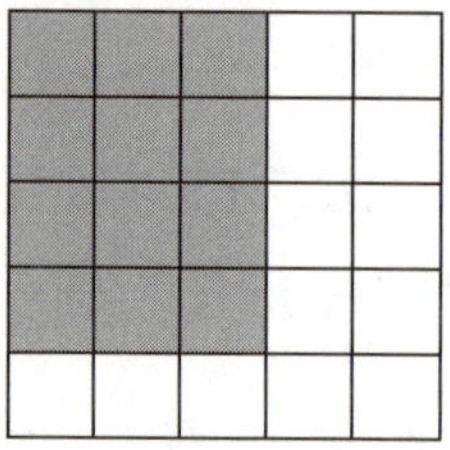

- ○ $6 \times 2 = 12$
- ○ $12 \times 1 = 12$
- ○ $4 \times 3 = 12$
- ○ $2 \times 6 = 12$

8 Barasa knits scarves. He knits 2 scarves a day for 3 days. How many scarves does Barasa knit in all?

0	1	5	6
○	○	○	○

9 What is the product of 4×2?

0	2	5	8
○	○	○	○

10 Peta has 3 baskets of socks. There are 5 socks in each basket. How should Peta count the socks?

- ○ skip count by 2s
- ○ skip count by 4s
- ○ skip count by 5s
- ○ skip count by 10s

11 Which multiplication sentence does the array show?

- ○ $4 \times 2 = 6$
- ○ $2 \times 4 = 6$
- ○ $3 \times 5 = 8$
- ○ $2 \times 4 = 8$

12 What is the product of 0×1?

0	1	10	100
○	○	○	○

STOP

Introduction to Division

Home Connection

English

Dear Family,

Today our class started **Chapter 3, Introduction to Division.** In this chapter, I will learn about different division strategies and models. I will also learn how to divide by 1, 2, and 5.

Love, ________________

Spanish

Estimada familia:

Hoy en clase comenzamos el **Capítulo 3, Introducción a la división.** En este capítulo aprenderé sobre las diferentes estrategias y modelos para dividir. También aprenderé a dividir entre 1, 2 y 5.

Cariños, ________________

Help at Home

You can help your child by asking him or her to divide common household items into equal groups, such as cereal pieces, paper clips, or crayons. Ask your child how many objects are in each equal group.

Ayude en casa

Usted puede ayudar a su niño pidiéndole que divida objetos comunes en grupos iguales, como trozos de cereal, sujetadores de papel o crayones. Pregúntele a su niño cuántos objetos hay en cada uno de los grupos iguales.

Math Online Take the chapter Get Ready quiz at macmillanmh.com.

Name ______________________________

Get Ready

Subtract.

1
  18
− 6
☐

2
  36
− 9
☐

3
  24
− 8
☐

Write a multiplication sentence.

4 

____ × ____ = ____

5 

____ × ____ = ____

Circle each number you count on the number line.

6 Skip count by 4s.

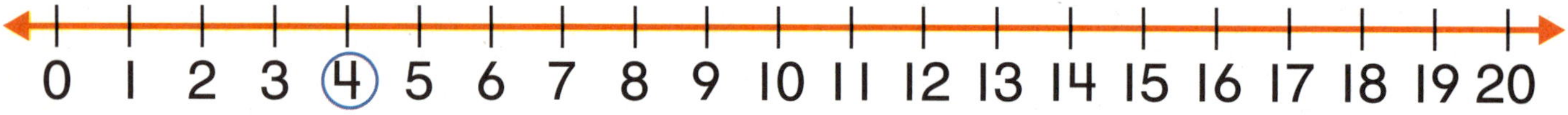

7 Skip Count by 6s.

Multiply.

8 $2 \times 8 =$ ____

9 $3 \times 7 =$ ____

10 $0 \times 9 =$ ____

STOP

Lesson 3-1

Name ________________________________

Relate Multiplication and Division

Key Concept

Multiplication and **division** are related.

There are 3 groups of counters.
There are 2 counters in each group.
There are $3 \times 2 = 6$ counters in all.

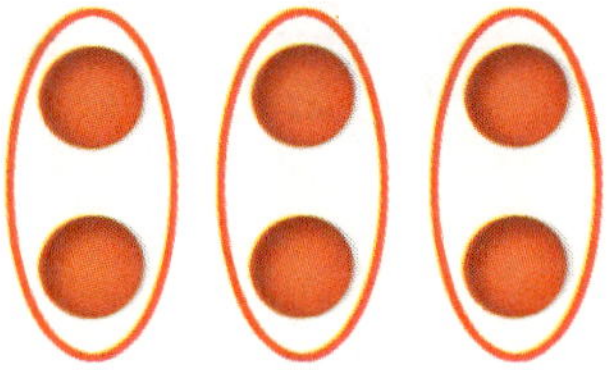

There are 6 counters in all.
They are divided into 3 equal groups of 2.
There are $6 \div 3 = 2$ counters in each group.

Vocabulary

multiply (multiplication) to combine equal groups

$$4 \times 2 = 8$$

Four groups of two are equal to eight or

$$2 + 2 + 2 + 2 = 8.$$

divide (division) to separate into equal groups

$$9 \div 3 = 3$$

A fact triangle shows the numbers in related multiplication and division sentences.

$3 \times 2 = 6$ $2 \times 3 = 6$

$6 \div 3 = 2$ $6 \div 2 = 3$

Example

Write related multiplication and division sentences.

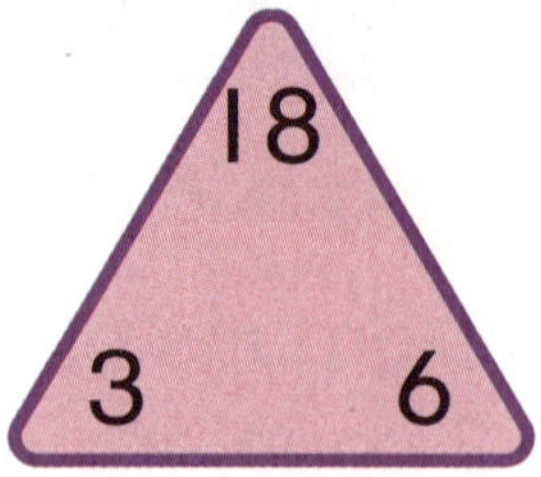

Step 1 There are 3 groups of frogs.
Step 2 There are 6 frogs in each group.
Step 3 There are 18 frogs in all.

Answer $3 \times 6 = 18$
$18 \div 3 = 6$

Step-by-Step Practice

Write related multiplication and division sentences.

Step 1 There are _____ groups of turtles.

Step 2 There are _____ turtles in each group.

Step 3 There are _____ turtles in all.

Answer _____ × _____ = _____

_____ ÷ _____ = _____

Name ______________________________

Guided Practice

Write related multiplication and division sentences.

1

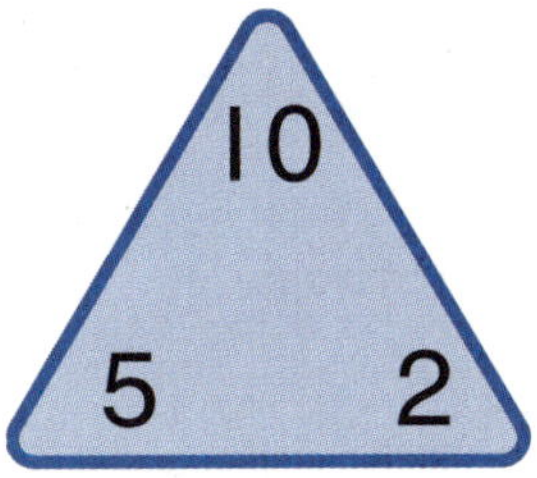

____ × ____ = ____

____ ÷ ____ = ____

2

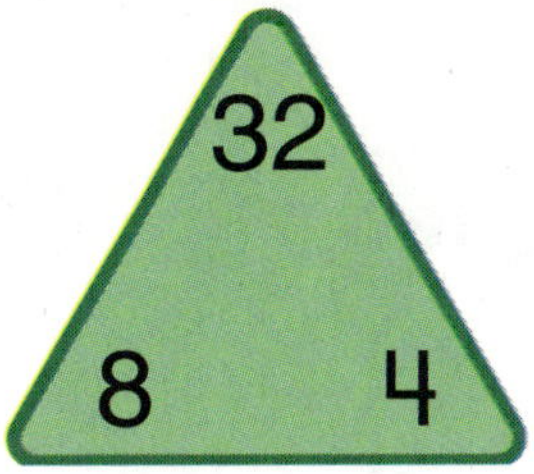

____ × ____ = ____

____ ÷ ____ = ____

Problem-Solving Practice

3 Riley has 14 shoes in her closet.
She sorted the shoes into pairs of 2.
How many pairs of shoes does Riley have?

Understand Underline key words.

Plan Draw a picture.

Solve Draw 14 shoes in groups of 2. Circle each group

Riley has ____ pairs of shoes.

Check Write related multiplication and division sentences.

____ × 2 = 14 14 ÷ ____ = 2

GO on

Practice on Your Own

Write related multiplication and division sentences.

4

____ × ____ = ____

____ ÷ ____ = ____

5

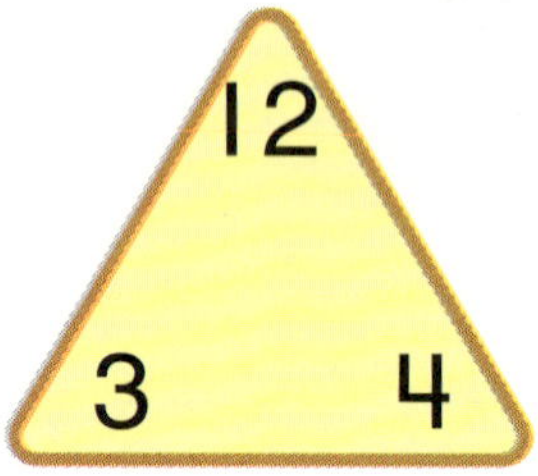

____ × ____ = ____

____ ÷ ____ = ____

6

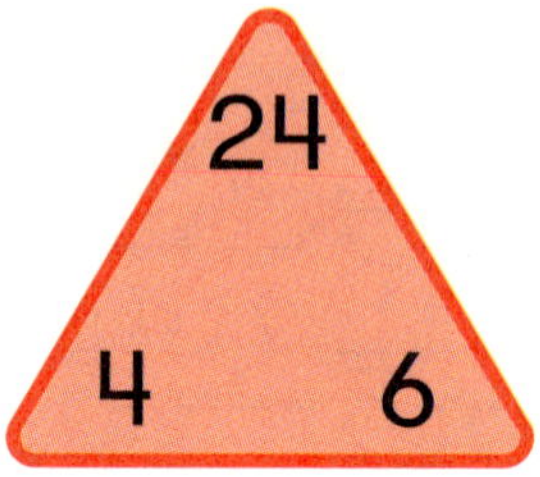

____ × ____ = ____

____ ÷ ____ = ____

7 **WRITING IN MATH** Luis plants 8 flowers in his garden. He makes 4 equal rows. How many flowers are in each row? Explain.

____

____

Vocabulary Check Complete.

8 You can use __________ to separate objects into equal groups.

STOP

Lesson 3-2

Name ______________________________

Repeated Subtraction and Skip Counting

Key Concept

How many equal groups of 5 can be made from 20?
The number 20 is shown on the number line.
Count back by 5s until you reach 0.

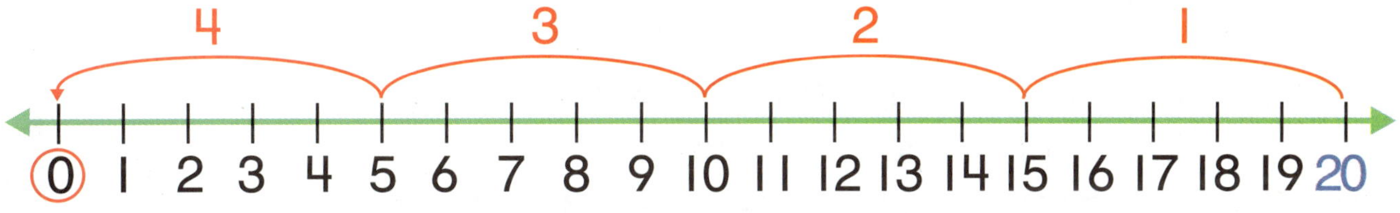

Subtract 5 from 20 until you reach 0.

$$\begin{array}{r}20\\-\ 5\\\hline 15\end{array}\quad\begin{array}{r}15\\-\ 5\\\hline 10\end{array}\quad\begin{array}{r}10\\-\ 5\\\hline 5\end{array}\quad\begin{array}{r}5\\-\ 5\\\hline 0\end{array}$$

$$20 \div 5 = 4$$

Vocabulary

divide (division) to separate into equal groups

repeated subtraction subtract the same number more than one time

$$15 - 5 - 5 - 5 = 0$$

skip count count objects in equal groups of 2 or more

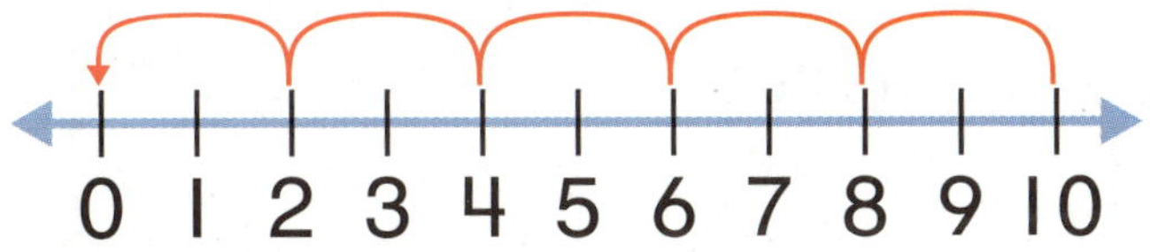

Example

There are 18 students in Mr. Lazo's class.
He wants to divide his students into teams of 3.
How many teams will he have?

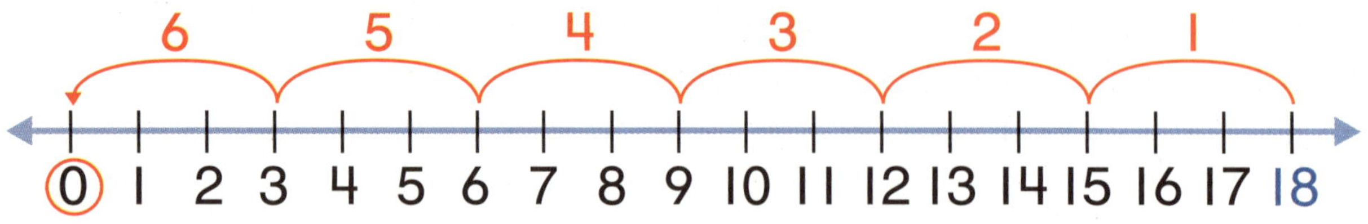

Step 1 Start at 18. Skip count backward by 3s.

Step 2 Subtract 6 times to reach 0.

Step 3 Write the division sentence. $18 \div 3 = 6$

Answer Mr. Lazo will have 6 teams.

Step-by-Step Practice

Molly has 16 toy cars.
She wants to divide the cars onto 4 shelves.
How many cars will be on each shelf?

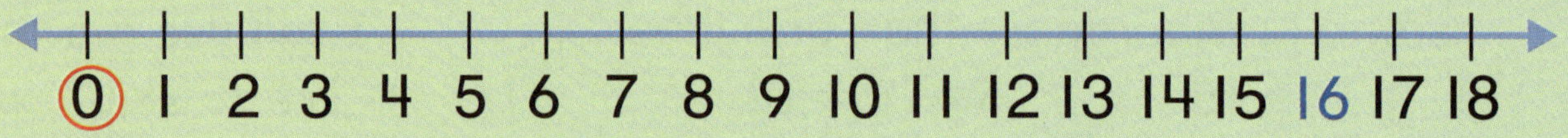

Step 1 Start at 16. Skip count backward by 4 s.

Step 2 Subtract _____ times to reach 0.

Step 3 Write the division sentence.

_____ ÷ _____ = _____.

Answer There will be _____ cars on each shelf.

Name ______________________________

Guided Practice

1 Divide. Use the number line.

21 ÷ 7 = ____

Problem-Solving Practice

2 Maria bought 25 prizes for her birthday party.
She gave 5 prizes to each friend.
How many friends were at Maria's party?

Understand Underline key words.

Plan Use a number line.

Solve Divide. 25 ÷ 5 = ____

Check Use a related multiplication sentence.

____ × 5 = 25

GO on

Practice on Your Own

3 Divide. Use the number line.

$14 \div 2 =$ ____

4 Write a division sentence. Use the number line.

____ $\div$ ____ $=$ ____

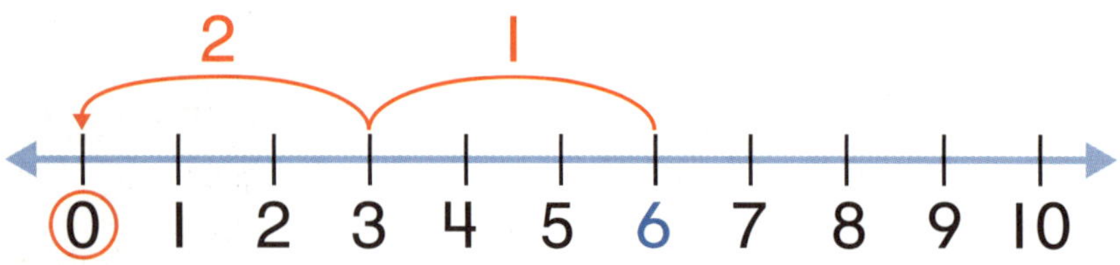

5 **WRITING IN MATH** Alberto has 8 books. He reads 2 books a week. How many weeks will it take Alberto to read all 8 books? How do you know?

____

____

____

Vocabulary Check Complete.

6 When you ______________________, you count objects in equal groups of 2 or more.

STOP

Name ______________________________

Progress Check 1 (Lessons 3-1 and 3-2)

Write related multiplication and division sentences.

1

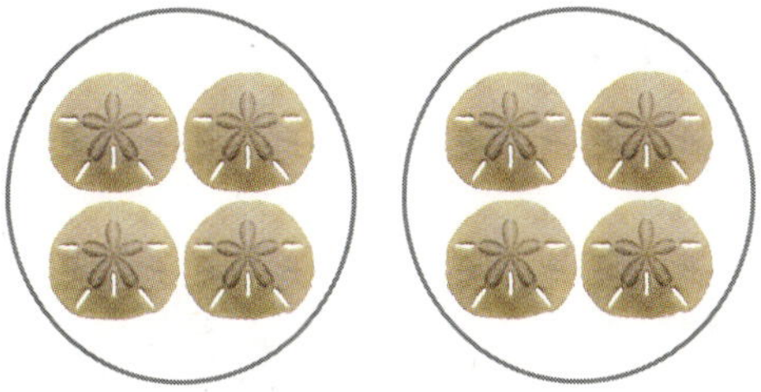

____ × ____ = ____

____ ÷ ____ = ____

2

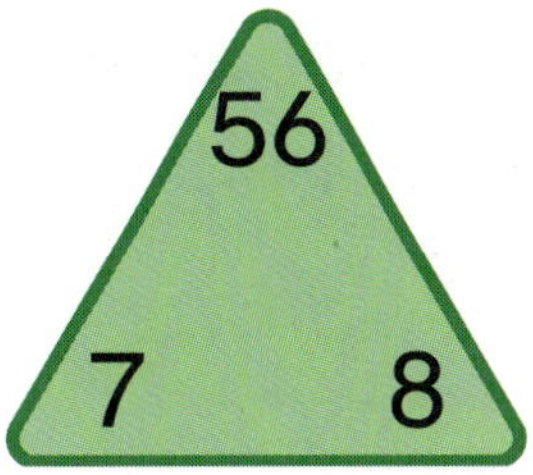

____ × ____ = ____

____ ÷ ____ = ____

3 Divide. Use the number line.

20 ÷ 2 = ____

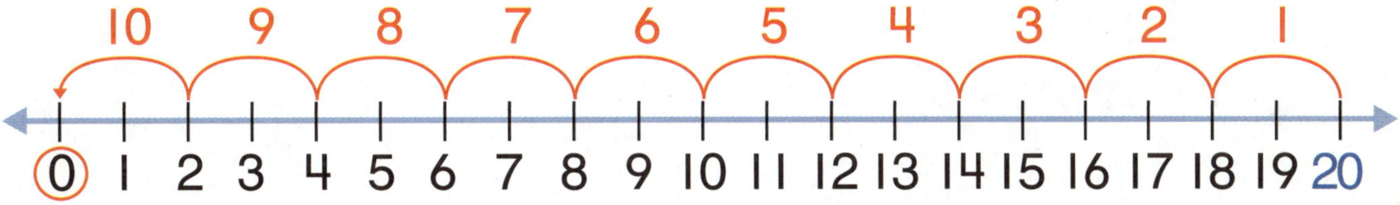

4 Rafael scored 12 goals during a soccer season. He scored 2 goals in every game he played. How many games did Rafael play this season?

Rafael played ____ games this season.

Name ______________________________

«« Replay Felix's Fun Facts!

Help Felix the Frog leap across the pond!

Find the missing number in each fact triangle below. Use each answer to help you find the missing number in the next triangle.

Lesson 3-3

Name ______________________________

Use Arrays to Model Division

Key Concept

You can use an **array** to divide.

There are 12 counters in all.
There are 3 rows.
There are 4 counters in each row.

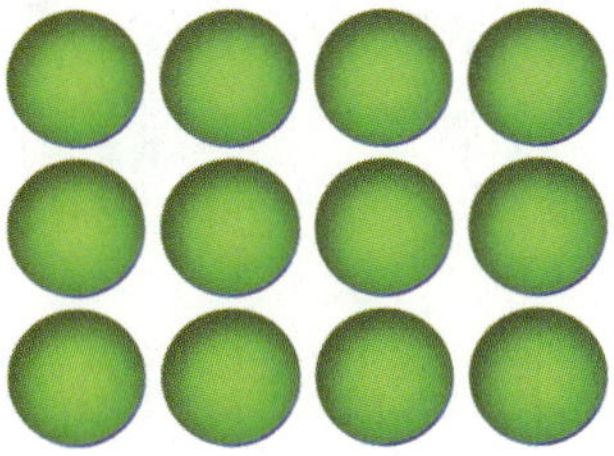

12	÷	3	=	4
↑		↑		↑
number of counters		number of rows		number of columns

Vocabulary

array objects displayed in rows and columns

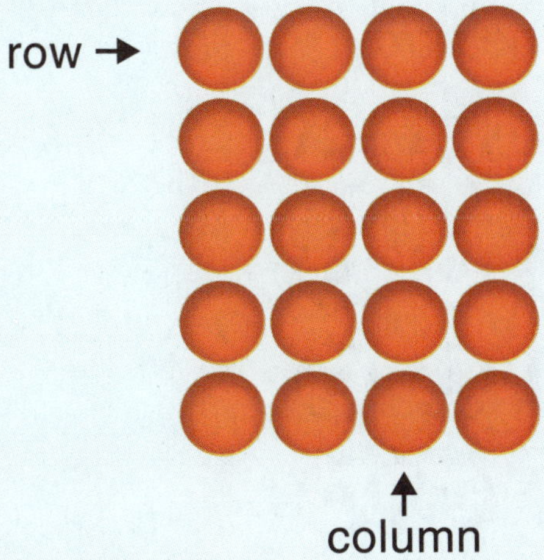

To make an array, put objects into equal rows and columns.

Example

Write a division sentence for the array.

Step 1 There are 15 beach balls in all.

Step 2 There are 3 rows.

Step 3 There are 5 columns.

Answer $15 \div 3 = 5$

Step-by-Step Practice

Write a division sentence for the array.

Step 1 There are 16 soccer balls in all.

Step 2 There are ____ rows.

Step 3 There are ____ columns.

Answer ____ ÷ ____ = ____

Name ______________________________

Guided Practice

Write a division sentence for each array.

1

___ ÷ ___ = ___

2

___ ÷ ___ = ___

Problem-Solving Practice

3 Akili works in a grocery store.
He put 18 pineapples in 2 equal rows.
How many pineapples are in each row?

Understand Underline key words.

Plan Draw a picture.

Solve Draw an array of 18 pineapples in 2 equal rows.

There are ___ pineapples in each row.

Check Count the number of columns in the array.

GO on

Practice on Your Own

Write a division sentence for each array.

4 

____ ÷ ____ = ____

5 

____ ÷ ____ = ____

6 Draw an array that shows $16 \div 2 = 8$.

7 An array has 12 balloons in 6 equal rows.

How many balloons are in each row? ____ balloons

8 WRITING IN MATH Keith put 14 stickers in 2 equal rows. How many stickers are in each row? Explain.

____

____

Vocabulary Check Complete.

9 Objects displayed in rows and columns are called

________________.

STOP

Lesson 3-4

Name ________________

Use Area Models to Show Division

Key Concept

You can use an **area** model to divide. $30 \div 6 =$ ____

30	$\div$	6	$=$	5
↑ number of squares		↑ number of rows		↑ number in each row

There are 30 squares.
There are 6 rows.
There are 5 squares in each row.
$30 \div 6 = 5$

An area model is similar to an array. Both have equal rows and equal columns.

Vocabulary

area the number of square units needed to cover the inside of an object

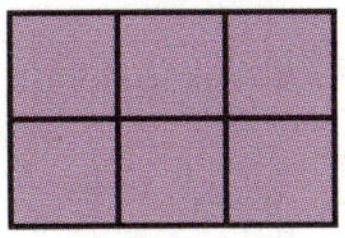

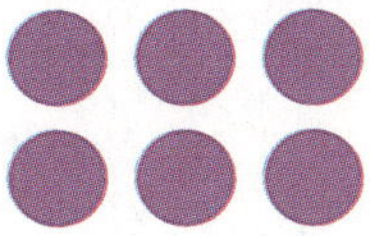

Example

Write a division sentence.

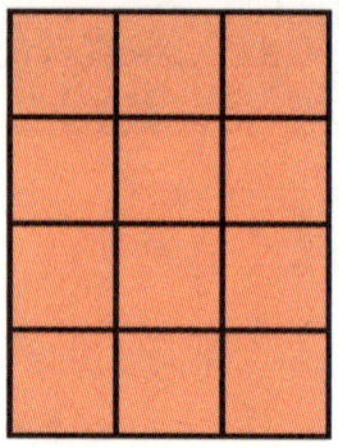

Step 1 There are 12 squares in all.

Step 2 There are 4 rows.

Step 3 There are 3 columns.

Step 4 The model shows 12 squares divided into 4 equal rows equals 3 columns of squares.

Answer $12 \div 4 = 3$

Step-by-Step Practice

Write a division sentence.

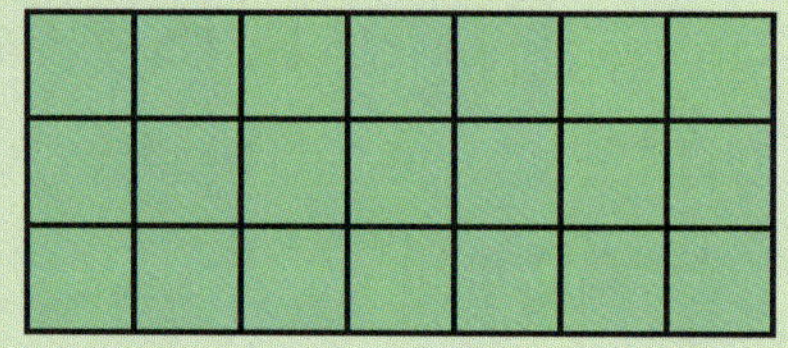

Step 1 There are 21 squares in all.

Step 2 There are ____ rows.

Step 3 There are ____ columns.

Step 4 The model shows ____ squares divided into ____ equal rows equals ____ columns of squares.

Answer ____ ÷ ____ = ____

Name ______________________________

Guided Practice

Write a division sentence.

1 ____ squares in all

____ rows and ____ columns

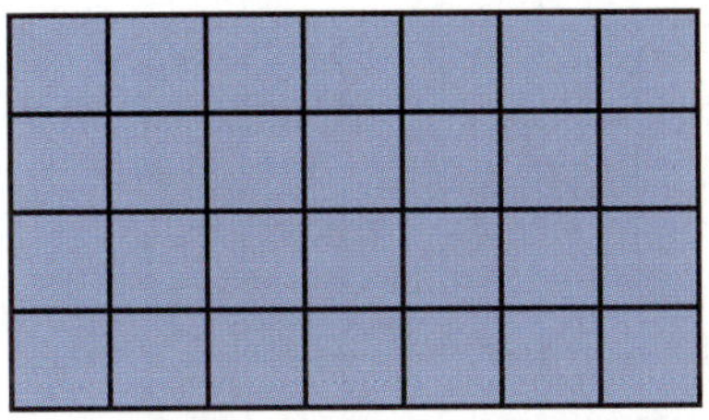

____ ÷ ____ = ____

2 ____ squares in all

____ rows and ____ columns

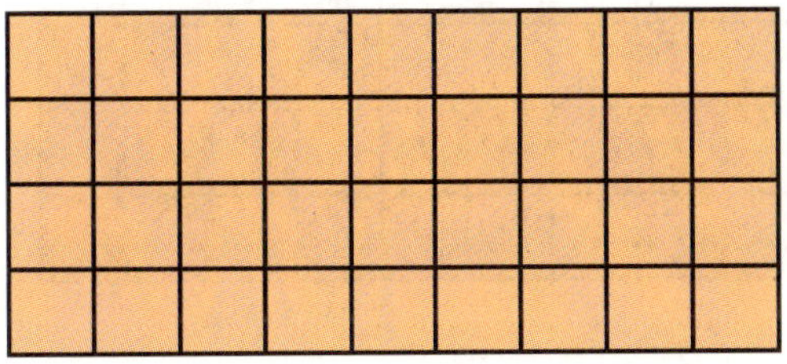

____ ÷ ____ = ____

Problem-Solving Practice

3 Caroline covers a rectangle with 30 tiles. She places 3 tiles in each row. How many rows of tiles are made?

Understand Underline key words.

Plan Draw an area model.

Solve Draw rows of ____ tiles.
Add rows until there

are ____ tiles in all.

Caroline makes ____ rows of tiles.

Check Did you draw 30 tiles in equal rows and columns?

GO on

Practice on Your Own

Write a division sentence.

4 ____ squares in all

____ rows and ____ columns

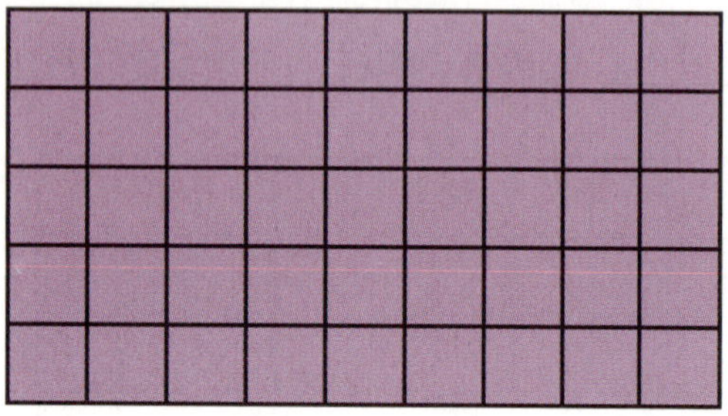

____ ÷ ____ = ____

5 ____ squares in all

____ rows and ____ columns

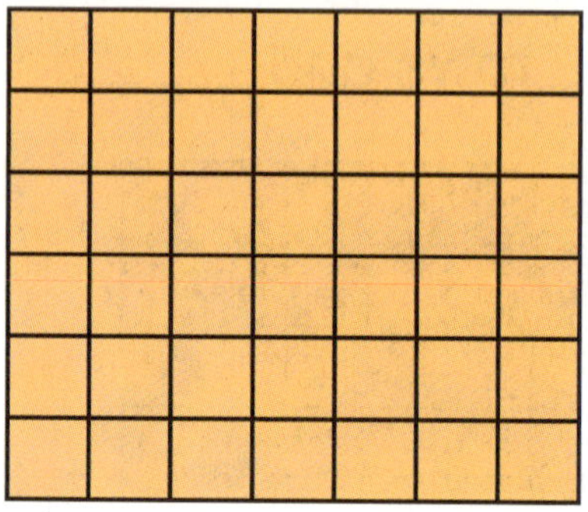

____ ÷ ____ = ____

6 An area model has 18 squares in all. There are 2 equal rows of squares.

____ ÷ ____ = ____

7 **WRITING IN MATH** Thomas places 64 tiles in equal rows of 8. How many rows of tiles does Thomas make? How do you know?

____

____

____

Vocabulary Check Complete.

8 ____________ is the number of square units needed to cover the inside of an object.

STOP

Name ______________________________

Progress Check 2 (Lessons 3-3 and 3-4)

1 Write a division sentence.

____ ÷ ____ = ____

2 Draw an array that matches the division sentence $9 \div 3 = 3$.

3 Write a division sentence for the area model.

____ ÷ ____ = ____

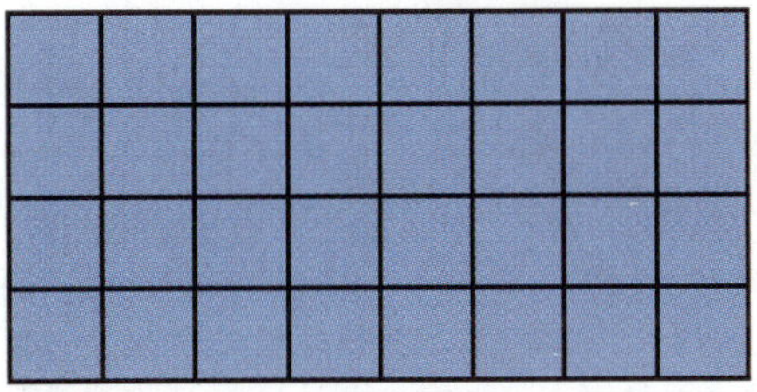

4 An area model has 25 squares in 5 equal rows.

How many squares are in each row? ____ squares

5 Kwasi arranged 20 recorders for music class in 2 equal rows. How many recorders did Kwasi place in each row? Explain.

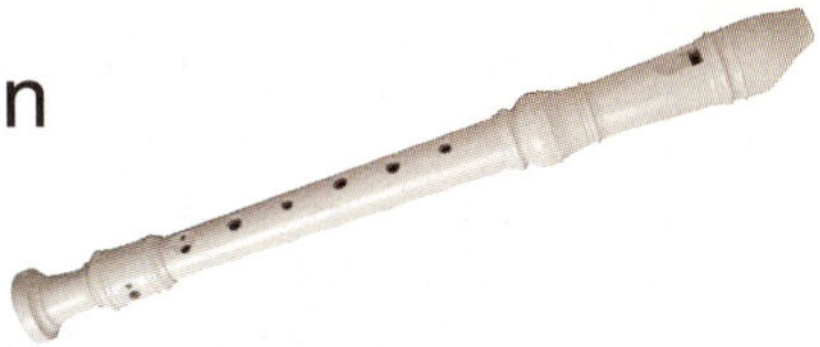

____

____

____

Name ______________________________

«« Replay

Complete each division sentence. Fill in the letters below to find the answer to the knock-knock joke.

C

____ ÷ ____ = ☐

P

____ ÷ ____ = ☐

M

____ ÷ ____ = ☐

A

____ ÷ ____ = ☐

O $24 \div 4 =$ ☐

L $18 \div 2 =$ ☐

☐ ☐ n ☐ e ☐ ☐ ☐ e ☐ ☐ ☐ y?

4 5 8 6 3 4 6 2 3 7 9 5

Lesson 3-5

Name ______________________________

Divide with 0 and 1

Key Concept

When 0 is divided by any number (except 0), the answer is 0.

0 objects divided into 4 groups

$0 \div 4 = 0$

The circles are empty because 0 objects are in each group.

When any number is divided by 1, the answer is that number.

3 crayons divided into 1 group

$3 \div 1 = 3$

Vocabulary

divide (division) to separate into equal groups

$9 \div 3 = 3$

3 strawberries divided into 3 plates equals 1 strawberry on each plate.

$3 \div 3 = 1$

Example

Write a division sentence.

Step 1 There are 12 flowers in all.

Step 2 The flowers are in 1 group.

Step 3 The picture shows 12 flowers divided into 1 group equals 12 flowers in each group.

Answer $12 \div 1 = 12$

Step-by-Step Practice

Write a division sentence.

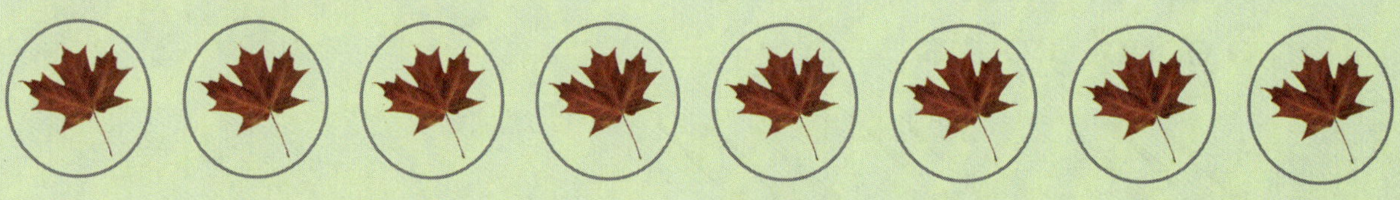

Step 1 There are __8__ leaves in all.

Step 2 The leaves are in ____ groups.

Step 3 The picture shows ____ leaves divided into ____ groups equals ____ leaf in each group.

Answer ____ ÷ ____ = ____

Name ______________________

Guided Practice

Write a division sentence.

1 0 crayons in 5 boxes

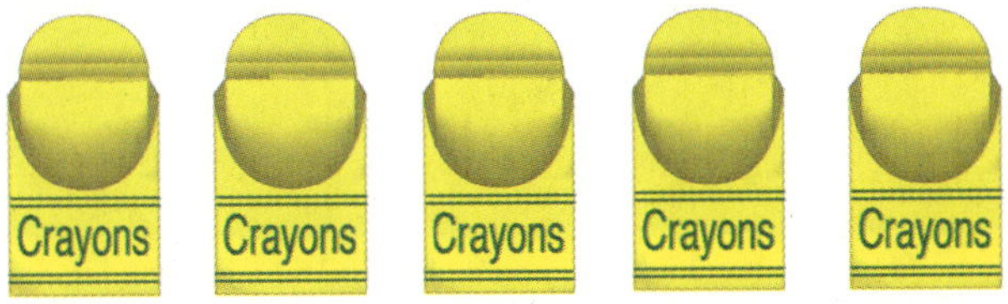

_____ ÷ _____ = _____

2 15 books on 1 shelf

_____ ÷ _____ = _____

Problem-Solving Practice

3 Juana has 12 fish. If she has only 1 fish tank, how many fish are in Juana's tank?

Understand Underline key words.

Plan Use logical reasoning.

Solve Juana has _____ fish in all. She will keep the fish in _____ group. The rule says any number divided by _____ is ______________.

_____ ÷ _____ = _____

There are _____ fish in Juana's tank.

Check Draw a picture to check your answer.

GO on

Practice on Your Own

Write a division sentence.

4 26 buttons in 1 group

____ ÷ ____ = ____

5 0 cookies on 2 plates

____ ÷ ____ = ____

6 Draw a picture for 6 ÷ 6 = 1.

Divide.

7 32 ÷ 1 = ____

8 45 ÷ 45 = ____

9 0 ÷ 19 = ____

10 0 ÷ 11 = ____

11 **WRITING IN MATH** Mrs. Romero has 25 markers. She divides her markers equally among 25 students. How many markers does each student have? How do you know?

____

____

____

Vocabulary Check Complete.

12 When you __________ a number by 1, the answer is that number.

STOP

Lesson 3-6

Name ______________________________

Divide by 2 and 5

Key Concept

A number line can be used to **divide** 20 by 5.

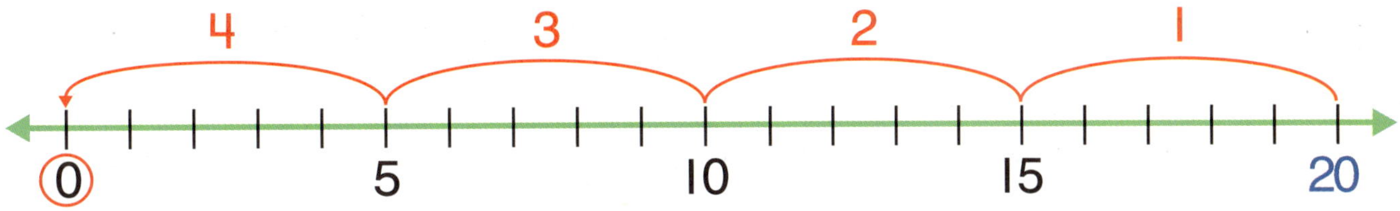

Skip count backward by 5s. Start at 20.
20, 15, 10, 5, 0 skip 4 times to reach 0.
$20 \div 5 = 4$

Vocabulary

divide (division) to separate into equal groups

$9 \div 3 = 3$

You can use multiplication to check division.

$25 \div 5 = 5$
$5 \times 5 = 25$

Example

Margaret has 30 connecting cubes. She places the cubes in 6 equal rows. How many cubes are in each row?

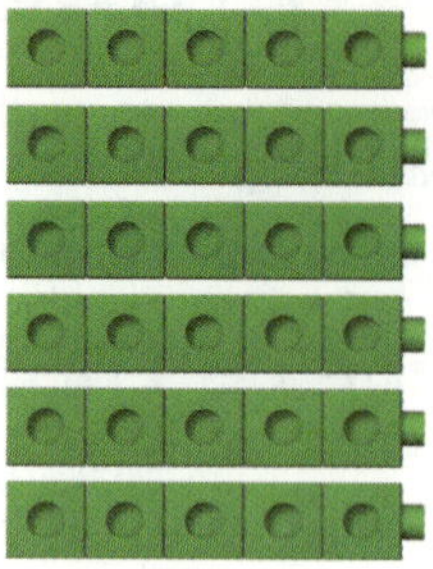

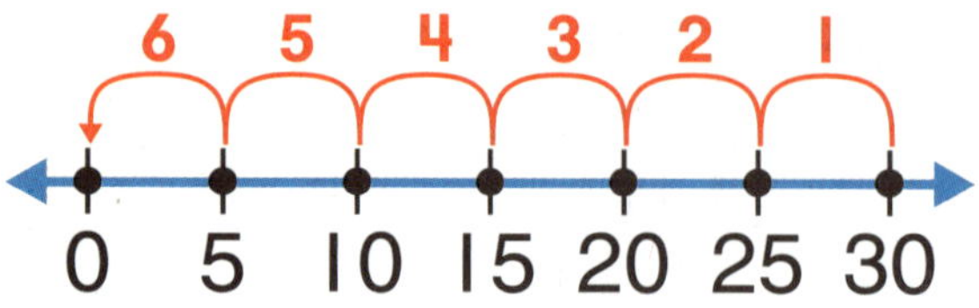

Step 1 Skip count backward by 5s. Start at 30.

Step 2 Skip 6 times. 30, 25, 20, 15, 10, 5, 0

Step 3 The division sentence is $30 \div 5 = 6$.

Answer There are 5 cubes in each row.

Step-by-Step Practice

Alonso bought a paint set. There are 12 paint colors in all. There are 2 rows of paint colors. How many paint colors are in each row?

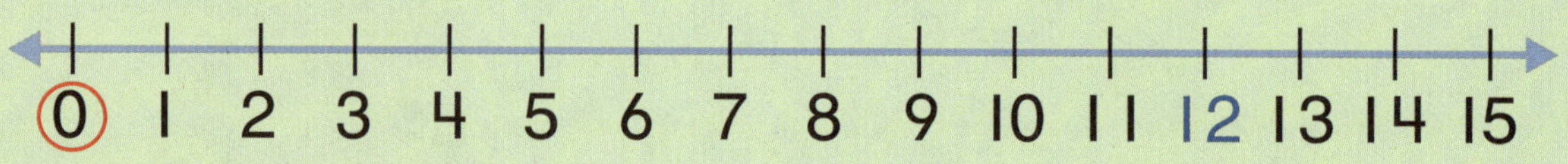

Step 1 Skip count backward by ____s. Start at ____.

Step 2 Skip ____ times.

____, ____, ____, ____, ____, ____, ____

Step 3 The division sentence is ____ ÷ ____ = ____.

Answer There are ____ paint colors in each row.

Name ______________________________

Guided Practice

1 Ian divides 15 marbles into 5 equal groups. How many marbles does Ian place in each group?

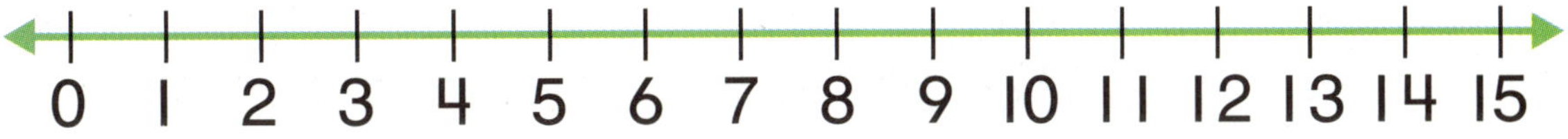

_____ ÷ _____ = _____ Ian places _____ marbles in each group.

Problem-Solving Practice

2 There are 35 students in Mr. Lin's music class. He places the students into 5 sections. How many students are in each section?

Understand Underline key words.

Plan Use a number line.

Solve Start at _____. Skip count backward by 5s until you reach _____ .

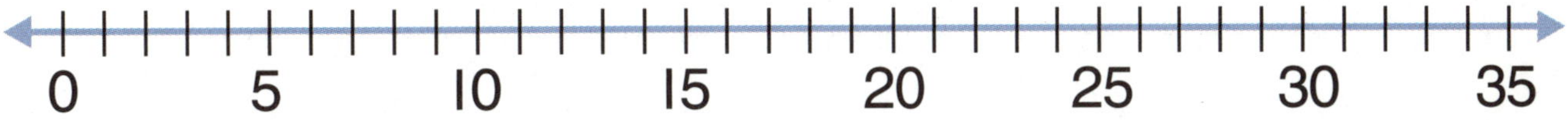

There are _____ students in each section.

Check Use a related multiplication fact.

GO on

Practice on Your Own

3 Kay makes an array of 20 tiles in 4 rows. How many tiles does Kay place in each row?

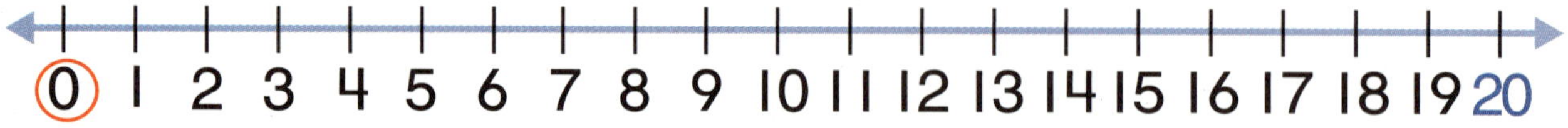

_____ ÷ _____ = _____ Kay places _____ tiles in each row.

4 Skip count backward to divide.

45 ÷ 5 = _____

____

5 WRITING IN ▶MATH There are 14 cowboy hats in a store. The hats are displayed in 2 groups. How many hats are in each group? Explain.

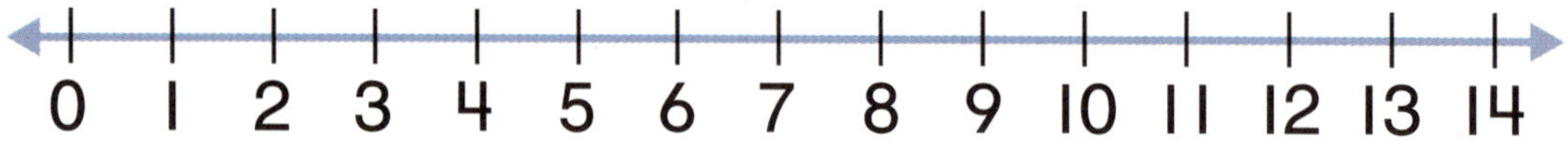

____

____

Vocabulary Check Complete.

6 In this lesson, we learned how to _______________ objects into equal groups of 2 or 5.

STOP

Name ______________________________

Progress Check 3 (Lessons 3-5 and 3-6)

1 Write a division sentence for the picture.

____ ÷ ____ = ____

2 Draw a picture for $18 \div 1 = 18$.

3 Meg bought 23 stamps. She wants to glue all of the stamps on 1 page of a notebook. How many stamps will be on the page?

____ ÷ ____ = ____

____ stamps

Divide.

4 $40 \div 5 =$ ____

5 $4 \div 2 =$ ____

6 $26 \div 26 =$ ____

7 $50 \div 5 =$ ____

8 Kevin has a package of 6 juice boxes. There are 2 equal rows of boxes. How many juice boxes are in each row? Explain.

____

____

Name ______________________________

« « Replay

What is the Mystery Picture?

Solve the division problems in the picture below.
Color the section **orange** when the answer is **0**.
Color the section **blue** when the answer is **any other number**.

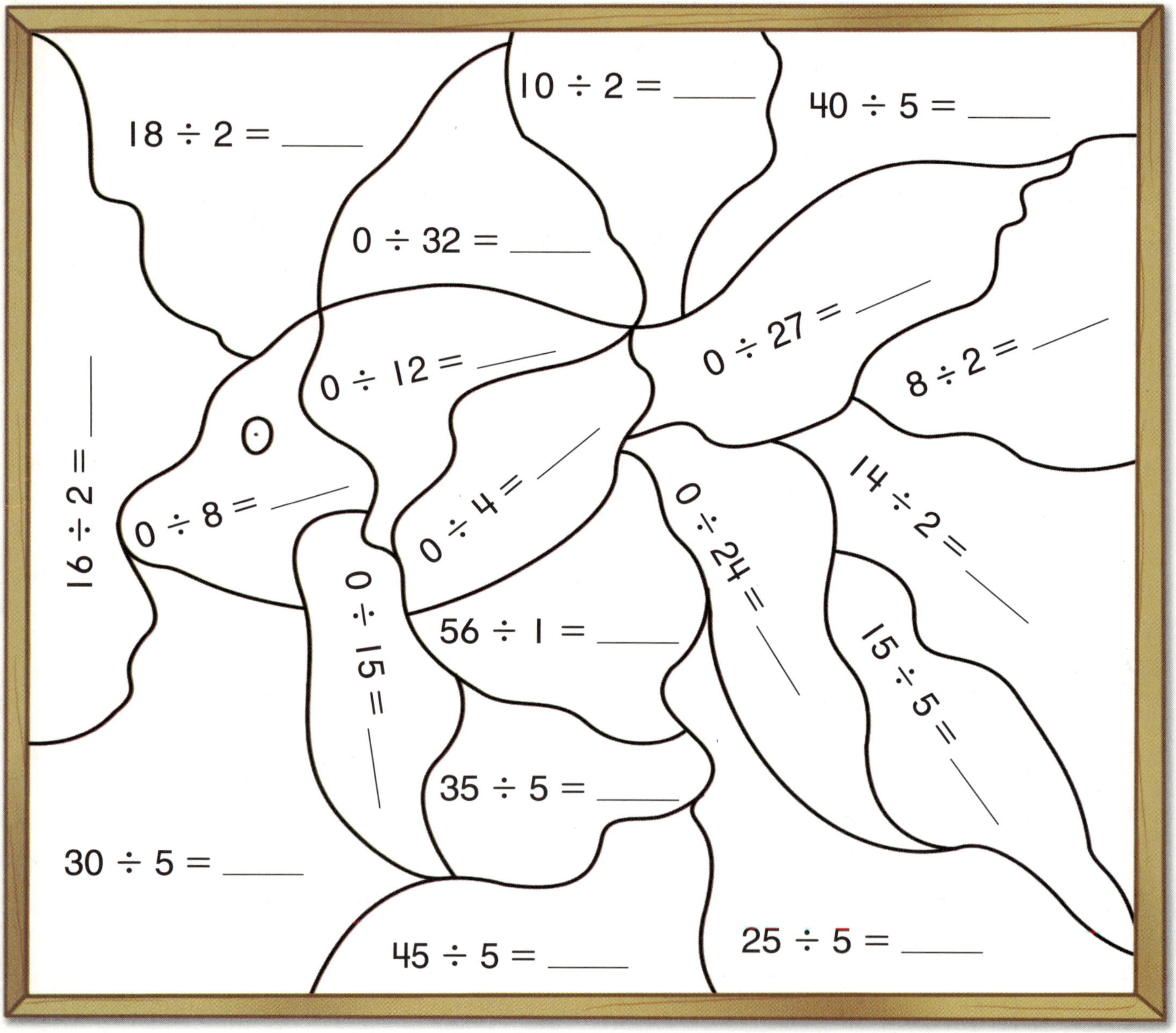

Name ______________________________

Review

Vocabulary

Word Bank

- area model
- array
- skip count

Use the Word Bank to complete.

1 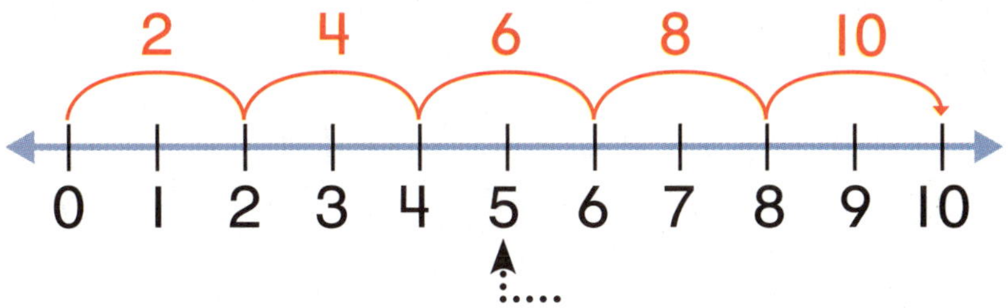

____

2 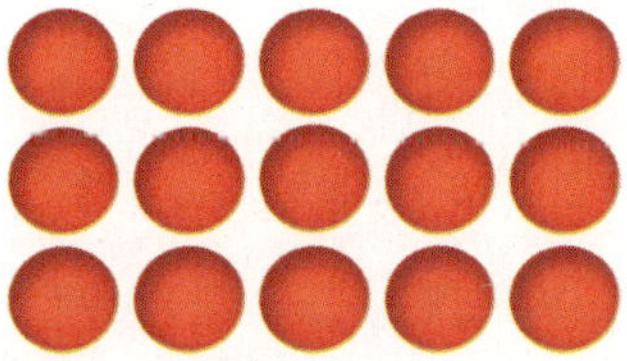

____

3 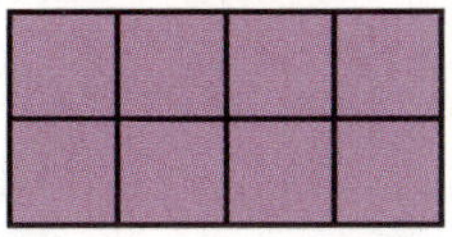

____

Concepts

Write related multiplication and division sentences.

4 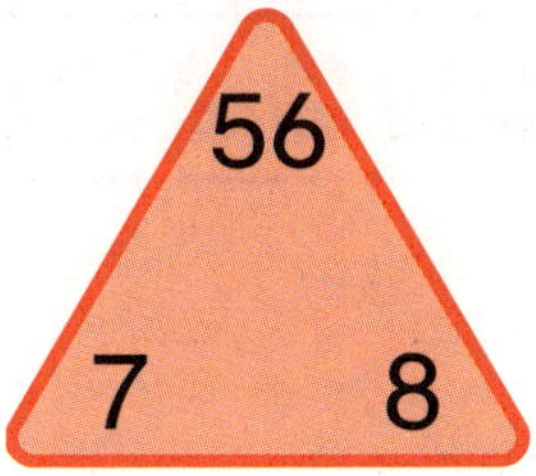

7 × 8 = ____

____ ÷ ____ = ____

5 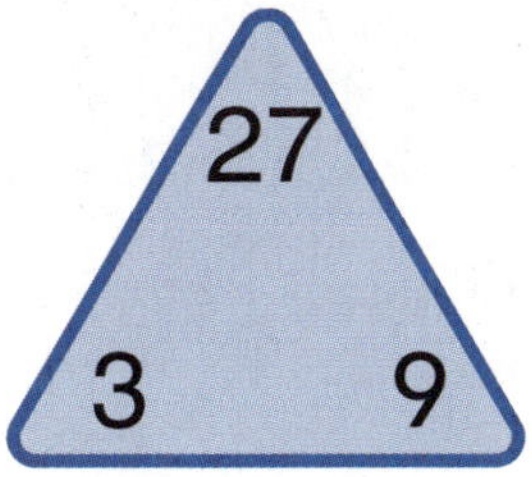

____ × ____ = ____

____ ÷ ____ = ____

GO on

Divide.

6

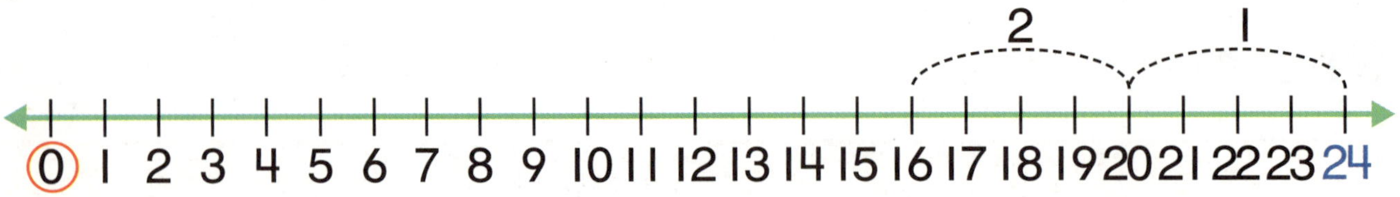

24 ÷ 4 = ____

7

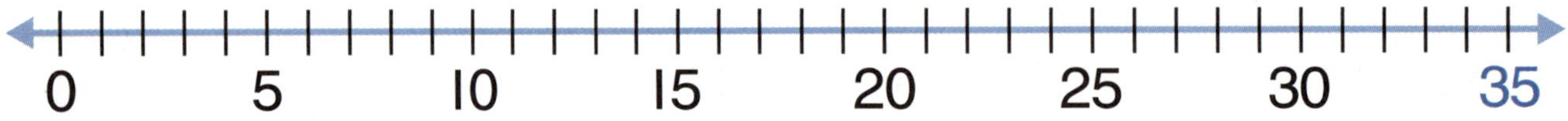

____ ÷ ____ = ____

Write a division sentence.

8 16 ÷ 4 = ____

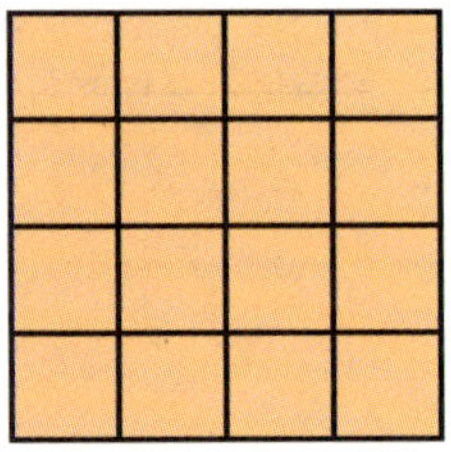

9 ____ ÷ ____ = ____

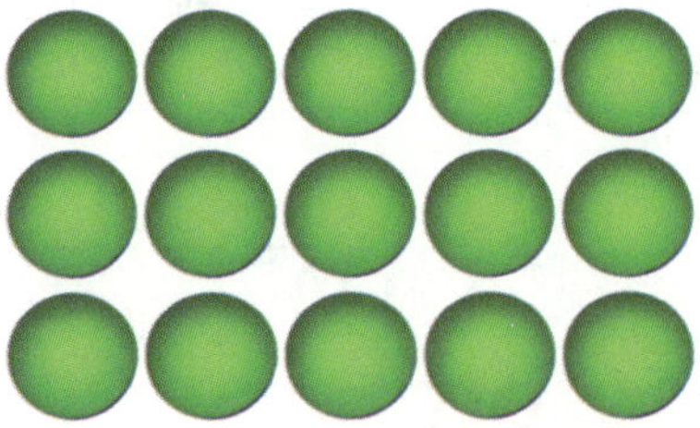

10 8 empty water glasses

0 ÷ 8 = ____

11 10 shells in 1 group

____ ÷ ____ = ____

Divide.

12 12 ÷ 2 = ____

13 10 ÷ 5 = ____

14 18 ÷ 2 = ____

STOP

Name ______________________________

Chapter Test

1 Write related multiplication and division sentences.

____ × ____ = ____

____ ÷ ____ = ____

2 Divide.

20 ÷ 4 = ____

Write a division sentence.

3 ____ ÷ ____ = ____

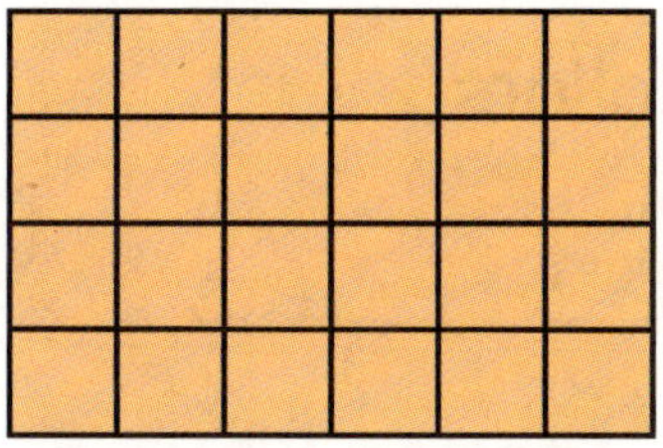

4 ____ ÷ ____ = ____

5 Draw a picture for 8 ÷ 1 = 8.

6 Who is Correct?

Juanita and Matthew solve $0 \div 12 =$ ____.

Juanita   Matthew

Circle the correct answer. Explain.

____

____

7 Miss Peta owns a bike shop.
She has 16 wheels in stock.
Each bike needs 2 wheels.
How many bikes can she make?

____ ÷ ____ = ____ Miss Peta can make ____ bikes.

8 Sarah went to a flower store. She has 25 dollars.
Potted flowers cost 5 dollars each.
How many potted flowers can Sarah buy?

____ ÷ ____ = ____ Sarah can buy ____ potted flowers.

Divide.

9 $14 \div 2 =$ ____

10 $30 \div 5 =$ ____

11 $0 \div 18 =$ ____

12 $13 \div 13 =$ ____

STOP

Name ______________________________

Test Practice

Choose the correct answer.

1 Which division fact is related to $6 \times 3 = 18$?

- ○ $6 \div 3 = 2$
- ○ $18 \div 9 = 2$
- ○ $6 \div 2 = 3$
- ○ $18 \div 3 = 6$

2 Tegene and Lilla divide 16 strawberries into 2 equal groups. How many strawberries are in each group?

4	8	12	14
○	○	○	○

3 Which division sentence matches this array?

- ○ $8 \div 2 = 4$
- ○ $5 \div 1 = 5$
- ○ $10 \div 2 = 5$
- ○ $2 \times 5 = 10$

4 Which division sentence is true?

- ○ $25 \div 5 = 5$
- ○ $25 \div 5 = 20$
- ○ $25 \div 5 = 25$
- ○ $25 \div 5 = 30$

5 Which division sentence matches this area model?

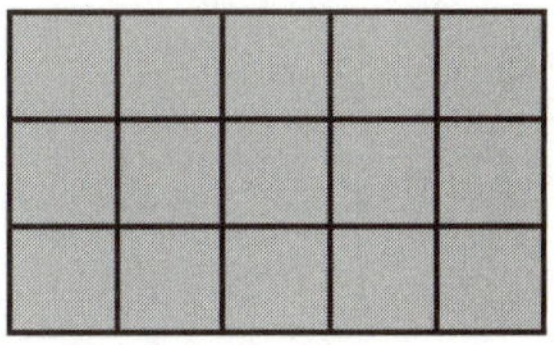

- ○ $15 \div 3 = 5$
- ○ $5 \times 3 = 15$
- ○ $15 \div 1 = 15$
- ○ $3 \times 5 = 15$

6 Which division sentence is shown by this picture?

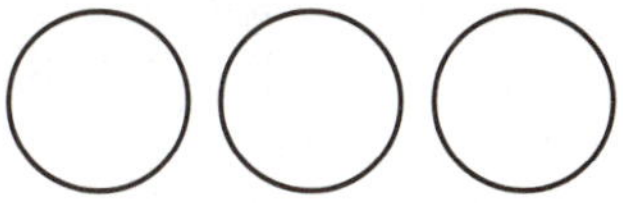

- ○ $0 \times 3 = 0$
- ○ $0 \div 3 = 0$
- ○ $3 \div 3 = 1$
- ○ $3 \div 1 = 3$

GO ON

7 Craig has 27 marbles. He wants to divide his marbles into 3 equal piles. How many marbles will be in each pile?

○ 9
○ 12
○ 19
○ 24

8 Which division sentence is true?

○ $14 \div 2 = 7$
○ $14 \div 2 = 12$
○ $14 \div 2 = 16$
○ $14 \div 2 = 28$

9 Which division sentence matches this array?

○ $12 \div 3 = 4$
○ $12 \div 2 = 6$
○ $12 \div 1 = 12$
○ $3 \times 4 = 12$

10 Which division sentence is shown by this picture?

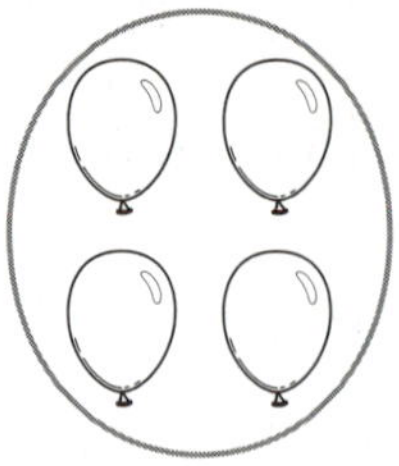

○ $0 \div 4 = 0$
○ $4 \div 2 = 2$
○ $4 \div 1 = 4$
○ $1 \times 4 = 4$

11 Which multiplication fact is related to $24 \div 8 = 3$?

○ $3 \times 6 = 18$
○ $6 \times 4 = 24$
○ $8 \times 3 = 24$
○ $24 \times 3 = 74$

12 Which division sentence matches this area model?

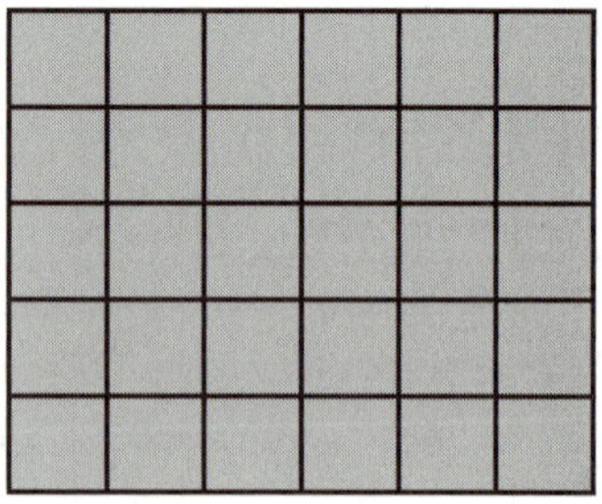

○ $5 \div 1 = 5$
○ $30 \div 5 = 6$
○ $30 \div 3 = 10$
○ $6 \times 5 = 30$

STOP